AI AND EXPERTISE:
Heuristic Search, Inference Engines, Automatic Proving

ELLIS HORWOOD SERIES IN ARTIFICIAL INTELLIGENCE
Joint Series Editors: Professor JOHN CAMPBELL, Department of Computer Science, University College London, and
Dr JEAN HAYES MICHIE, Knowledgelink Limited, Edinburgh

* In preparation

AI AND EXPERTISE:
Heuristic Search, Inference Engines, Automatic Proving

HENRI FARRENY
Professor of Computer Science
Institut National Polytechnique de Toulouse
E.N.S.E.E.I.H.T., Department of Computer Science, France

Translator:
JONATHAN BARCHAN
Intasoft Limited, Exeter

ELLIS HORWOOD LIMITED
Publishers · Chichester

Halsted Press: a division of
JOHN WILEY & SONS
New York · Chichester · Brisbane · Toronto

This English edition first published in 1989 by
ELLIS HORWOOD LIMITED
Market Cross House, Cooper Street,
Chichester, West Sussex, PO19 1EB, England
The publisher's colophon is reproduced from James Gillison's drawing of the ancient Market Cross, Chichester.

Distributors:

Australia and New Zealand:
JACARANDA WILEY LIMITED
GPO Box 859, Brisbane, Queensland 4001, Australia

Canada:
JOHN WILEY & SONS CANADA LIMITED
22 Worcester Road, Rexdale, Ontario, Canada

Europe and Africa:
JOHN WILEY & SONS LIMITED
Baffins Lane, Chichester, West Sussex, England

North and South America and the rest of the world:
Halsted Press: a division of
JOHN WILEY & SONS
605 Third Avenue, New York, NY 10158, USA

This English edition is translated from the original French edition *Exercices pro-grammes d'intelligence artificielle,* published in 1987 by Masson Editeur, Paris, © the copyright holders.

British Library Cataloguing in Publication Data
Farreny, Henri
AI and expertise: heristic search, inference engines, automatic proving. —
(Ellis Horwood series in artificial intelligence).
1. Problem solving. Applications of artificial intelligence
I. Title   II. Exercices programme's d'intelligence artificielle. *English*
006.3
Library of Congress Card No. 88–38081

ISBN 0–7458–0426–8 (Ellis Horwood Limited)
ISBN 0–470–21384–1 (Halsted Press)

Typeset in Times by Ellis Horwood Limited
Printed in Great Britian by Hartnolls, Bodmin

Though already full of energy, our science is currently in its infancy, yet it is portrayed as though it were ancient; its stammerings evoke the repulsion of senility instead of the joy which youth spreads in its eager efforts.

In any case, the principal goal of teaching must be to convey the notion of the living and on-going effort which science makes to adapt to the realities of the outside world, to build the harmonious edifice of our representation from the principles or hypotheses decreed by the mind guided by experimental reasoning. At the same time, instead of encumbering the future technician with the useless baggage of unconnected formulae and facts we place a useful tool in his hand, one which, and I must insist on this point, is easy to repair; once he has seen how it is made he will know how to maintain it for himself in the light of scientific progress, having been accustomed for so long to the idea that we do not possess a definitive system.

Paul Langevin
La Pensee et l'Action,
Editions Sociales 1950
(written in 1904)

To Mariví, Angela and Henri
for their patience and affection

¡ Ay ! Ya viene la madrugada,
Los gallos estan cantando.
Compadre, estan anunciando
Que ya empieza la jornada . . . Ay . . . Ay . . .

¡ Ay ! Al vaivén de mi carreta,
Nació esta lamentación.
Compadre, ponga atención
Que ya empieza mi cuarteta.
No tenemos protección . . . Ay . . . Ay . . .

¡BASTA YA ! (Atahualpa Yupanqui)

Oh! Dawn is approaching
The cocks are singing
My friend, they are announcing
That our day is starting . . . Oh . . . Oh . . .

Oh! From the reeling of my cart
comes this lamentation.
My friend, pay attention,
my couplet will begin.
We are living without protection . . .

Table of contents

Introduction

In many training courses the participants are asked to complete 'projects', which consist of implementing and testing basic methods, or variants, for solving practical problems. Thinking about the subject matter of a project and then actually putting it into practice helps in the assimilation and deeper understanding of the course. This book is designed to extend a foundation course in (i) heuristic problem-solving using state spaces, (ii) expert system generators, and (iii) automatic theorem proving. In particular it may accompany (FARRENY & GHALLAB 1986).

This book is divided into three large chapters corresponding to each of these three themes.

Heuristic problem-solving by means of state spaces is traditionally one of the principal elements in general problem-solving techniques; although historically they were developed independently, the techniques combined under this label bring to mind certain approaches or preoccupations belonging to operational research and graph theory. Accounts are available which are directly of use for foundation studies, in particular: (NILSSON 1971), (NILSSON 1980), (LAURIERE 1986); for higher levels, see (PEARL 1985) or (FARRENY & GHALLAB 1986). Chapter 1 deals in particular with the implementation of type A, A^*, A^{e*} and $A^{\alpha*}$ algorithms. Following the presentation of general programs, several different ways of applying these methods to the problems of the travelling salesman and the 8-puzzle are compared.

Constructing expert systems is currently one of the most popular domains for applied AI. Many research laboratories are interested in this area. At present research efforts are especially concerned with knowledge representation and inference capabilities, algorithms, and operating environments.

Accounts that are of direct use to foundation studies are to be found in: (NILSSON 1980), (ALTY & COOMBS 1984), (FARRENY 1985), (WATERMAN 1985), (BENCHIMOL *et al.* 1986), (LAURIERE 1986), (FARRENY & GHALLAB 1986). The first part of chapter 2 shows the construction of a diagnostic-type expert system generator. The final version consists of an engine which is interactive, self-explaining, backward chaining, works depth first according to a tentative regime system, and reasons using uncertain knowledge in the style of **EMYCIN**; this engine can be asked (several types of) questions *during* the diagnostic session as well as *after* it;

furthermore, the user can correct his earlier replies *during* the session and then continue (according to various principles). The second part of the chapter shows the construction of a planning-type expert system generator. The final version has an engine which is nonmonotonic, uses mixed chaining, works depth first according to a tentative regime system with control over the depth, and searches for all solutions within the fixed depth. With a view to tackling the construction of generators which allow variables, the third part shows the development of a relatively sophisticated pattern matching system.

Automatic theorem proving is another important branch of problem-solving. The best-known product from this body of work is the **PROLOG** language, initially developed at the University of Marseille: **PROLOG** interpreters are a particular kind of theorem prover. One of the very rare French AI books to have appeared before 1980 deals with automatic theorem proving: (PITRAT 1970). Accounts which are directly usable in foundation studies are to be found in: (NILSSON 1980), (DELAHAYE 1986), (LAURIERE 1986), (FARRENY & GHALLAB 1986). For deeper investigation, see: (CHANG *et al.* 1973), (LOVELAND 1978), (SIEK-MANN 1983), (WOS *et al.* 1984). The first part of chapter 3 shows the construction of a Prolog-type prover — hence intended for Horn clauses — which follows a depth first linear input strategy. The final version searches for and displays all proofs (which accord with **PROLOG**'s strategy); it accepts predefined predicates, starting with cut (/) and evaluable functions. In the second part of the chapter another theorem prover is built which is still limited to Horn clauses but follows a positive unit strategy driven breadth first, with elimination of subsumed unit clauses. The two demonstrators are tested and perfected on a variety of examples.

Each of the three chapters contains substantial review of the ideas which are presumed to have been studied beforehand. The basic algorithms are described in detail. They are all written in **LISP**, specifically the **Le-Lisp dialect, version 15.2** (CHAILLOUX 1986). The programs are usually introduced in successive stages; for example, there are 5 stages in the construction of a **PROLOG**-type theorem prover. The progressive nature of the explanation should allow the reader to master fully the relatively complex programs which are obtained by the end of each chapter. The working of each program is illustrated by reproducing actual usage sessions; all of the programs have been implemented and tested on an Apple Macintosh Plus micro with a megabyte of main memory. The explanation is interspersed with notes intended to complement the general principles presumed to have been acquired in an earlier course — these notes should help in a tangible manner because they are in the context of a practical objective.

A rudimentary knowledge of **LISP** will suffice to tackle this book; however, an inexperienced **LISP** programmer would be advised to follow the chapters in the order given, while a moderately experienced one may launch into the three chapters in any order. Deliberately, only a very small number of the primitive functions in **LISP** are used (Le-Lisp 15.2 offers several hundred) and the programming style adopted aims (in principle!) at intelligibility for the largest number of readers, rather than conciseness or efficiency of execution; this constraint is — somewhat — relaxed as the chapters progress.

The majority of Le-Lisp 15.2 primitives which are used have the same definition in the most common LISP dialects. Some primitives (probably very few: TAG, EXIT for instance) may take other names, obey other conventions, or even not be defined in the dialect you possess. To check the relationships, the reader can refer to (CHAILLOUX *et al.* 1986), or, failing that, to: (GIRARDOT 1985), (ROY & KIREMITDJIAN 1985) or (WERTZ 1985), all recent works in which Le-Lisp plays a central role.

Apart from the strict concern for intelligibility, the programming style is influenced by some personal customs which are perhaps debatable. For example, the programs given never contain IF or LET but frequently use COND and PROG (PROG will often be linked with calls to RETURN); in some limited circumstances variables which lie outside of a group of procedures are used; certain side effects (caused in particular by recourse to SETQ rather than LET) could perhaps be avoided without serious damage to readability.†

The algorithms provided prior to the programs in Le-Lisp are represented in a very simple code which conforms to current conventions. To follow the rest of this paragraph, the reader is recommended to take a brief look at the algorithms found at the start of chapters 1, 2, and 3. The symbol $<--$ represents the assignment operation: to the identifier which precedes it, from the value of the expression which follows it. The main control primitives have the classical forms: UNTIL ... REPEAT ... ENDUNTIL, IF ... THEN ... ENDIF, IF ... THEN ... ELSE ... ENDELSE. The STOP primitive causes the algorithm to terminate. The keyword PROCEDURE comes before the name of a sequence of instructions grouped into a procedure; any procedure which is called must return a value; the names of one or more formal parameters which will be treated as variables local to the procedure may appear, in parentheses, after the name of a procedure. The keyword VARIABLE (or VARIABLES) precedes the declarations of local variables which are in addition to the formal parameters of procedures. The RETURN primitive causes exit from the procedure where it is located, the value returned by the procedure being that of the expression which follows RETURN; RETURN is always used to exit a procedure. Reading and writing are caused by READ and WRITE.

---

† Nonetheless, the author shares the opinion of principle expressed by Douglas Hofstadter: 'I personally am not a fanatic about avoiding **setq's** and other functions that cause side effects. Though I find the applicative style to be just-elegant, I find it impractical when it comes to the construction of large AI-style programs' (HOFSTADTER 1983).

Quan surts a fer el viatge cap a Itaca
has de pregar que el camí sigui llarg,
ple d'aventures, ple de coneixenca.
Has de pregar que el camí sigui llarg,
que siguin moltes les matinades,
que entrarás en un port que els teus ulls ignoraven,
que vagis a ciutats a aprendre dels que saben.
Tingues sempre al cor la idea d'Itaca

ITACA (Kavafis — Lluis LLach)

When you undertake the voyage to Ithaca
pray that the way be long,
full of adventures, full of knowledge.
Pray that the way be long,
that there be many mornings
when you enter a port unknown to your eyes
that you go into cities to learn from those who know.
Keep always in your heart the idea of Ithaca

1

Heuristic resolution by state spaces

The representation and resolution of problems by means of state spaces constitutes one of the main branches of general AI techniques. The first section is concerned with a review of the definition of the fundamental algorithms (A, A^*, A^{e^*}, A^{α^*} algorithms) and their properties. Thereafter, the principal aim will be to implement algorithms of type A and A^* for application to 'the travelling salesman problem' and 'the 8-puzzle'. For basic explanations refer to: [NILSSON 1971], [NILSSON 1980], [PEARL 1984], [LAURIERE 1986] and [FARRENY & GHALLAB 1986].

1.1 SEARCH ALGORITHMS IN STATE SPACES

1.1.1 Notion of state spaces
In a great many cases it is useful to represent a statement of a problem in terms of *states* and *state transformation operators*. Take as an example the problem from the game best known as 8-puzzle given in Fig. 1.1 below.

```
    2 8 3          ?            1 2 3
    1 6 4       --- >          8 / 4
    7 / 5                       7 6 5

      P1                         P2
```

Fig. 1.1 — An 8-puzzle problem.

The / symbol marks the empty square. Any square which is adjacent horizontally or vertically to the empty one (3 squares in position P1) can slide into its place; this is a basic move in the puzzle. A sequence of basic moves is sought which will turn position P1 into position P2. Three moves are available from P1: **6 downwards, 7 rightwards, 5 leftwards**. We can deduce a priori that there are 4 moves per occupied square, giving 32 possible moves. It is best to consider only the 4 equivalent moves which can be made by the empty square: **rightwards, leftwards, upwards, downwards**. Thus the problem can be seen as perfectly defined by the data of the *initial state* represented by means of position P1, the *goal state* represented by means of position P2 and the 4 *state transformation operators* (represented by: **empty square rightwards, leftwards, upwards, downwards**, an operator only being usable if the empty square is not touching an edge prohibiting the move). By applying the operators to the initial state, 3 new *states* for the board may be produced, and so on step by step.

Let us associate a node with each state. Create an arc (m,n) each time that a state n can be obtained by applying a state transformation operator to m (node n is a *child* of m). The graph obtained is called *a state space* of the problem under consideration. This graph may not be finite in the general case. The node s corresponding to the initial state is a *root* of the graph; if n is the node which stands for any state, at least one *path* from s to n exists. The number of arcs on this path constitutes *a distance from* s *to* n. However, the state space is not necessarily a *tree*: more than one path may exist from s to n. Generally a state space is considered to have just one root, i.e. there is only one initial state for the problem: if that were not the case, one could get back to the previous situation by creating an initial super-state of which all the initial states were children. From now onwards we shall fuse every state with the node which corresponds to it in the state space being considered.

We call the set of children of s *depth level* 1. *Depth level* p (p>1) encompasses all the nodes which possess at least one parent in depth level p−1 but which have no parent at a level less than p−1. Every node in level p lies at minimum distance p from the initial state.

Hypothesis 1.1
We shall suppose from now on that the number of children belonging to any state is finite, but unbounded. Given this, for any integer p, the set of nodes at a distance less than p from the initial state is finite, and thus the set of nodes in the state space is enumerable.

1.1.2   Progressive development of a state space
Given a state space representing a particular problem, the solution to the problem can be viewed as the search for a path between the initial state and a goal state within the state space. But for practical reasons (memory, processing time) we are not able or do not want actually to construct the whole of the state space: it is only a matter of constructing a sub-graph of the state space containing a solution-path. This sub-graph is called a *search graph*. *Search algorithms in state spaces* differ according to the way they direct or constrain the growth of the search graph. They periodically test whether a goal state lies in the current search graph. If this is not the case, they choose a node n in the search graph and apply state transformation operators to it; by this route children of n may be added to the current search graph. The operation which produces all the child-states of a given state by applying all of the available operators is termed *complete development of the state*; certain algorithms apply only a part of the available operators: this is *partial development*. From now on, unless expressly noted, complete developments will be applied (and we shall refer simply to *development*).

Algorithm 1.1 below (which we shall call an *unordered search* algorithm in contrast to the *ordered search* algorithm presented later on) represents a large class of search algorithms. This algorithm is *nondeterministic* because the way in which the choice indicated in instruction 5 is made is not specified. Instructions 16 to 26 correspond to the development of state m.

For an explanation of the general notation used when supplying algorithms refer to the information in the foreword.

Algorithm 1.1: Unordered search

```
1        head <-- {s}
2        store <-- { }
3        father(s) <-- "nothing"
4        UNTIL head = { } REPEAT
5            m <-- some element of head
6            IF m is a goal THEN
7                UNTIL m = "nothing" REPEAT
8                    display m
9                    m <-- father(m)
10               ENDUNTIL
11               STOP
12           ENDIF
13           head <-- head - {m}
14           store <-- store - {m}
15           children <-- set of children of m
16           UNTIL children = { } REPEAT
17               n <-- some element of children
18               children <-- children - {n}
19               father(n) <-- m
20               IF n belongs to store THEN
21                   head <-- head + {n}
22                   store <-- store - {n}
23               ENDIF
24               IF n belongs to neither head nor store THEN
25                   head <-- head + {n}
26               ENDIF
27           ENDUNTIL
28       ENDUNTIL
```

The { } symbol denotes the empty set. Executing the STOP instruction (instruction 11) causes the algorithm to terminate. It may also stop as a result of satisfying the test head={ } in instruction 4. The symbol s represents the initial state; head and store are sets; father can be thought of as referring to a one-dimensional array. It should be noted that, in the general case, there is no guarantee that algorithm 1.1 will terminate, and that if it does terminate there is no guarantee that a goal-state will have been reached (even if a goal state does actually exist in the state space).

1.1.3 Ordered search algorithms

In certain applications, for each state n there may exist a *numeric evaluation ff(n)* to measure *the promise of the state* in terms of its presence on a solution path: in a choice between several nodes in the search graph, the development of one of the most promising states will be preferred (from a *heuristic* point of view).

The function termed ff is called an *evaluation function*. From now on we shall suppose that the values returned by this function are positive and that the smaller they are, the more promising the states.

Consider algorithm 1.2 on the next page. This algorithm is a particular example of algorithm 1.1: the existence of the evaluation function ff is used to guide the choices of the old instruction 5. This kind of algorithm is known as *ordered search algorithm* or *best first search algorithm*.

Here father and value can be seen as identifiers referring to one-dimensional arrays; ff stands for the evaluation function which is defined for each state n. The algorithm can be terminated by instruction 12 or by normal exit from the UNTIL loop (instruction 5).

Algorithm 1.2: Ordered search

```
1       head <-- {s}
2       store <-- { }
3       father(s) <-- "nothing"
4       value(s) <-- ff(s)
5       UNTIL head = { } REPEAT
6           focus <-- set of m's in head for the smallest possible value(m)
7           IF a goal m exists in focus THEN
8               UNTIL m = "nothing" REPEAT
9                   display m
10                  m <-- father(m)
11              ENDUNTIL
12              STOP
13          ENDIF
14          m <-- some element of focus
15          head <-- head - {m}
16          store <-- store + {m}
17          children <-- set of children of m
18          UNTIL children = { } REPEAT
19              n <-- some element of children
20              children <-- children - {n}
21              IF n belongs to head and ff(n) ≤ value(n) THEN
22                  father(n) <-- m
23                  value(n) <-- ff(n)
24              ENDIF
25              IF n belongs to store and ff(n) ≤ value(n) THEN
26                  head <-- head + {n}
27                  store <-- store - {n}
28                  father(n) <-- m
29                  value(n) <-- ff(n)
30              ENDIF
31              IF n belongs to neither head nor store THEN
32                  head <-- head + {n}
33                  father(n) <-- m
34                  value(n) <-- ff(n)
35              ENDIF
36          ENDUNTIL
37      ENDUNTIL
```

As in the case of algorithm 1.1, there is no certainty that algorithm 1.2 will come to an end, nor that, if it does end, a goal state will have been found. Algorithm 1.2 limits the choices but is still nondeterministic as to precisely which state is to be developed.

Let us return to the 8-puzzle problem discussed in §1.1.1 (see Fig. 1.1). We shall define a state evaluation function, ff, as follows:

ff(m)=p if p filled squares in state m are not in the
place assigned to them in the goal-state

Fig. 1.2 above shows the search graph obtained after 6 state developments:

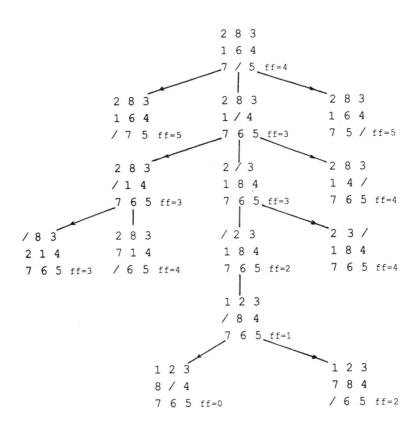

Fig. 1.2 — An ordered search algorithm applied to an 8-puzzle problem.

just as we prepare to set off the 7th development the goal-state is discovered. To implement algorithm 1.2, the choice in the focus set is made by giving preference to the state which appeared earliest (if necessary, the sibling nodes at the left of the figure are considered older than those at the right).

1.2   ALGORITHMS A, A^*, A^{e*}, $A^{\alpha*}$

In certain problems we may be able to associate a numeric value c(m,n) with each arc (m,n) in the state space. For example, c(m,n) may represent the *cost* of transforming state m into state n. The *cost of a solution path* may then be defined as the sum of the *basic costs* associated with the arcs of which it is composed.

So, for example, if the cost of each basic move of the 8-puzzle is assigned a value of one unit, the cost of the solution path found in Fig. 1.2 comes to 5.

Hypothesis 1.2
Henceforward only basic costs which are positive or null will be considered.

In certain cases, obtaining an *optimal solution path* (that is one which costs the least) is of the utmost interest. In other cases, the cost of the solution paths found is only relatively important or even unimportant. Nevertheless, regardless of the weight attached to minimising the costs of solution paths, the information relating to the costs of the arcs may be used to order the search.

Hypothesis 1.3
From now on we shall suppose that for any positive δ the number of paths in the state space whose cost is less than δ is finite. A sufficient condition for this hypothesis to be true is that there exists a strictly positive d such that all arc costs are greater than d. Another sufficient condition is that the state space is finite and has strictly positive basic costs.

1.2.1   A algorithms
By definition, these are governed by an ordered search type algorithm (see algorithm 1.2 earlier) using an evaluation function ff of the following kind:

$$\text{for any state m, } ff(m) = gg(m) + hh(m)$$

The term gg(m) represents *the cost of the shortest path currently known from the initial state s to state m*. Hence the function gg actually depends on two arguments: state m (an explicit argument) and the current stage reached in the algorithm. Note that the value of the shortest path from s to m may perhaps only appear when a state is developed (or redeveloped); to be more precise, we should say: gg(m,i), where i stands for the number of developments tried, and not simply: gg(m). In the end, given a search graph, the value of gg(m) will be completely defined for every state of the graph.

For any m, hypothesis 1.2 ensures that there is a shorter path in the state space from s to m. Let g(m) be the cost of such a path. Clearly: $gg(m) \geqslant g(m)$. gg(m) may thus be considered an *overestimate* of g(m).

The term hh(m) is simply constrained to depend on m alone. It is sometimes called *the heuristic term*, although the term gg(m) also plays a heuristic role as a part of the evaluation function used by algorithm 1.2. But the value of gg(m) is perfectly defined given the current search graph and the values of the arcs in this graph, while the same information is not generally sufficient to determine hh(m).

From now onwards, the term gg(m) will be called: *standard term*. The function hh will be called: *heuristic function*. In numerous problems hh(m) may be interpreted as an estimate of the cost of a minimum path from m to the closest of the goal states.

Algorithm 1.3: Ordered search of type A

```
1        head <-- {s}
2        store <-- { }
3        father(s) <-- "nothing"
4        gg(s) <-- 0
5        ff(s) <-- hh(s)
6        UNTIL head = { } REPEAT
7            focus <-- set of m's in head for the smallest possible ff(m)
8            IF a goal m exists in focus THEN
9                UNTIL m = "nothing" REPEAT
10                   display m
11                   m <-- father(m)
12               ENDUNTIL
13               STOP
14           ENDIF
15           m <-- some element of focus
16           head <-- head - {m}
17           store <-- store + {m}
18           children <-- set of children of m
19           UNTIL children = { } REPEAT
20               n <-- some element of children
21               children <-- children - {n}
22               IF n belongs to head and gg(m) + c(m,n) ≤ gg(n) THEN
23                   father(n) <-- m
24                   gg(n) <-- gg(m) + c(m,n)
25                   ff(n) <-- gg(n) + hh(n)
26               ENDIF
27               IF n belongs to store and gg(m) + c(m,n) ≤ gg(n) THEN
28                   head <-- head + {n}
29                   store <-- store - {n}
30                   father(n) <-- m
31                   gg(n) <-- gg(m) + c(m,n)
32                   ff(n) <-- gg(n) + hh(n)
33               ENDIF
34               IF n belongs to neither head nor store THEN
35                   head <-- head + {n}
36                   father(n) <-- m
37                   gg(n) <-- gg(m) + c(m,n)
38                   ff(n) <-- gg(n) + hh(n)
39               ENDIF
40           ENDUNTIL
41       ENDUNTIL
```

Algorithm 1.3 above is the result of putting the general search algorithm

1.2 specifically into an algorithm of type A. **father**, **gg** and **ff** can be regarded here as one-dimensional arrays; **hh** represents the actual heuristic function, defined for each state. Like its predecessors, this algorithm can terminate either by executing **STOP** (instruction 13) or by normal exit from the **UNTIL** loop (instruction 6).

Note that the way in which algorithm 1.3 is expressed keeps it nondeterministic with regard to which node of focus is chosen for development (see instruction 15). The nodes in focus might, for instance, be classified in order of appearance, or in relation to the value of the standard term, or in relation to the value of the heuristic term . . .

1.2.1.1   Particular A algorithms

Algorithms with a monotonic function hh

Function hh is said to be *monotonic on a state space E* when:

$$Vm,n \in E, \text{ with } m \text{ father of } n: hh(m) \leqslant hh(n) + c(m,n)$$

Monotonicity expresses a certain coherence between the values returned by hh and the costs associated with the arcs. Let E be a state space which obeys hypotheses 1.1 to 1.3 and let an A algorithm be applied to E, using a function hh which is monotonic on E. It can be proved that each time a state m is developed: $gg(m) = g(m)$ (minimum possible value of gg in m). The result is that if a state is ordered in **store** it is impossible for it to come out later on (and go into **head**). Hence algorithm 1.3 can be simplified by eliminating lines 27 to 33.

Uniform cost algorithm

When the function hh is always null, the algorithm obtained is known as a *uniform cost algorithm*. In fact an algorithm which is said to be of uniform cost is obtained once hh is always constant; the constant may be considered as a *uniform* estimate of the cost of a minimum path from any node to the nearest goal. The null function hh is monotonic, so lines 27 to 33 of algorithm 1.3 can be eliminated. In this last form, the uniform cost algorithm is close to that of Moore–Dijkstra, which is often taught in graph theory for finding optimal-cost paths; the essential difference between the two algorithms stems from the fact that one is concerned with guiding the progressive development of the graph (state space), while the other works on a graph which is available in its entirety from the outset.

Breadth first algorithm

When, in addition, the values of the arcs are uniformly equal to 1, we get an algorithm which belongs to the family of *breadth first* algorithms: a state belonging to depth-level p cannot be developed until all the states in levels $q < p$ have been developed. On account of the unitary values associated with the arcs, algorithm 1.3 can be simplified further in order to reproduce the classical expression of breadth first algorithms: these allow a state to be tested to see whether it is a goal at the moment it appears (as a result of developing another state) instead of when it is about to be developed.

1.2.1.2 Notable properties of A algorithms

It can easily be shown that on each pass through instruction 6 of algorithm 1.3 the current search graph is an *arborescence,* with the initial state as the root.

It may be proved that, if a state space contains at least one goal-state and satisfies hypotheses 1.1 to 1.3, any algorithm of type A applied to this state space will terminate on discovering a goal-state.

1.2.2 A* algorithms

Call $h(n)$ the cost of a minimum path, if one exists, between some state n and the closest goal-state. We notice that if hypothesis 1.3 is true and if a path between n and a goal exists, then a minimum path between n and the goals will exist. If no path exists between n and the goals, it will be the case that $h(n)=\infty$.

Algorithms of type A are called A* algorithms if they satisfy the additional constraint:

$$\forall n \ hh(n) \leqslant h(n)$$

Here $hh(n)$ appears as a reduced estimate of $h(n)$.

1.2.2.1 Particular A* algorithms

If hypothesis 1.2 is true, we have: $\forall n \ h(n) \geqslant 0$. In that case, the uniform cost algorithm (and hence the breadth first algorithm, obtained as a particular case) is an A* algorithm since:
$$\forall n \ hh(n) = 0 \leqslant h(n)$$

1.2.2.2 Notable properties of A* algorithms

It can be shown that, if a state space contains at least one goal-state and satisfies hypotheses 1.1 to 1.3, every algorithm of type A* which is applied to this state space will terminate on encountering a minimum cost path between the initial state and the set of goal-states. This result is called the *admissibility theorem.*

Let E be a state space, and @1 and @2 be two A* algorithms which can be applied to E. Let hh1 and hh2 be the respective heuristic functions of @1 and @2. Algorithm @2 is said to be *better informed* then @1 if:

$$\forall n, \text{ a non-goal state, } hh1(n) < hh2(n)$$

It can be shown that, if a state space E contains at least one goal-state and satisfies hypotheses 1.1 to 1.3, and if @1 and @2 are two A* algorithms applicable to E such that @2 is better informed than @1, then every state developed by @2 at the time it is applied to E is also developed by @1 when it is applied to E. Thus the set of states in the final search graph obtained by @2. is included in the set of states in the final search graph obtained by @1. This result will be termed the *comparison theorem*.

1.2.3 A^{e*} algorithms
Algorithms of type A are called A^{e*} algorithms if they satisfy the constraint:

$\forall n$ $hh(n) \leq h(n) + e$ where e is a constant

A^* algorithms are particular cases of A^{e*}.

Take a state space which contains a goal-state and which satisfies hypotheses 1.1 to 1.3. Let C_o be the cost of a minimum path from the initial state to the goal-states. It can be shown that any algorithm of type A^{e*} which is applied to this state space will terminate when it discovers a path (between the initial state and a goal-state) whose cost is less than or equal to $C_o + e$.

1.2.4 $A^{\alpha*}$ algorithms
Algorithms of type A are called $A^{\alpha*}$ algorithms if they satisfy the constraint:

$\forall n$ $hh(n) \leq (1 + \alpha)h(n)$ where α is a constant

A^* algorithms are particular cases of $A^{\alpha*}$.

Take a state space which contains a goal-state and which satisfies hypotheses 1.1 to 1.3. Let C_o be the cost of a minimum path from the initial state to the goal-states. It can be shown that any algorithm of type $A^{\alpha*}$ which is applied to this state space will terminate when it discovers a path (between the initial state and a goal-state) whose cost is less than or equal to $(1+\alpha)C_o$.

1.3 BASIC LISP PROGRAM FOR A ALGORITHMS (and A^*, A^{e*}, $A^{\alpha*}$)

Program 1.1 below turns algorithm 1.3 introduced in §1.2.1 into Le-Lisp (version 15.2). The main procedure is ordsrchA: **ord**ered **search** by **A** algorithms. Detailed comments are supplied at the end of the program.

Program 1.1

```
1        (DE ordsrchA ( )
2        (PROG (node head store choice)
3           (SETQ ndevels 0 ncreated 0
4                head (LIST (create (initialstate) NIL 0)))
                              ;initialstate: specific procedure
5           (UNTIL (NULL head)
6              (SETQ choice (focus))
7              (COND ((NOT (NUMBERP choice))
                         (RETURN (result choice))))
8              (transferfather)
9              (SETQ ndevels (1+ ndevels))
10             (orderchildren (CAR store)
                         (develop (GET (CAR store) 'state))))
                              ;develop: specific procedure
11          (PRINT "failure")))
```

```
1    (DE create (state father gg)   ;Called by: ordsrchA
2        (PROG (lab)
3            (SETQ ncreated (1+ ncreated) lab (GENSYM))
4            (PUTPROP lab state 'state)
5            (PUTPROP lab father 'father)
6            (PUTPROP lab gg 'gg)
7            (SET lab (+ gg (hh state)))
                            ;hh: specific procedure
8            (RETURN lab)))
```

```
1    (DE focus ( )   ;Called by: ordsrchA
2        (PROG (list ff)
3            (SETQ list head ff (EVAL (CAR head)))
4            (UNTIL (OR (NULL list) ( (EVAL (CAR list)) ff))
5                (COND ((goal? (GET (CAR list) 'state)) (RETURN (CAR list))))
                        ;goal?: specific procedure
6                (SETQ list (CDR list)))
7            (RETURN 1)))
```

```
1    (DE result (lab)   ;Called by: ordsrchA
2        (PROG ( )
3            (TERPRI)
            (PRINT "Here is one solution whose value is " (GET lab 'gg))
4            (TERPRI) (displaystate (GET lab 'state))
                        ;displaystate: specific procedure
5            (UNTIL (NULL (SETQ lab (GET lab 'father)))
6                (TERPRI) (PRINT "obtained from ")
7                (TERPRI) (displaystate (GET lab 'state)))))
```

```
1    (DE transferfather ( ) ;Called by ordsrchA
2        (SETQ store (CONS (CAR head) store) head (CDR head)))
```

```
1    (DE orderchildren (father childlist) ;Called by ordsrchA
2        (PROG (lab ff)
3            (UNTIL (NULL childlist)
4                (SETQ lab NIL)
5                (COND ( (AND (SETQ lab (exists? (CAR child-list) head))
6                (SETQ ff (ffreduced? lab father (CADR childlist))))
7                        (updatenode lab father (CADR child-list) ff)
8                        (reorghead lab ff))
9                    ((AND (NULL lab)
10                        (SETQ lab (exists? (CAR child-list) store))
11                        (SETQ ff (ffreduced? lab father (CADR
                            childlist))))
12                        (updatenode lab father (CADR child-list) ff)
13                        (addtohead lab ff))
14                    ((NULL lab)
15                        (SETQ lab
16                            (create (CAR childlist)
                                father (+ (GET father 'gg) (CADR
                                childlist))))
17                        (addtohead lab (EVAL lab))))
18                (SETQ childlist (CDDR childlist)))))
```

```
1   (DE exists? (state listlab)   ;Called by: orderchildren
2       (PROG ( )
3           (UNTIL (NULL listlab)
4               (COND ( (EQUAL state (GET (CAR listlab) 'state
5                       (RETURN (CAR listlab))))
6               (SETQ listlab (CDR listlab)))
7           (RETURN NIL))

1   (DE ffreduced? (child father cost) ;Called by: orderchildren
2       (PROG (ff)
3           (COND ((< (SETQ ff (+ (− (EVAL child) (GET child 'gg))
                                    (GET father 'gg) cost))
4                   (EVAL child))
5               (RETURN ff)))
6           (RETURN NIL)))

1   (DE updatenode (child father cost ff)   ;Called by: orderchildren
2       (PUTPROP child father 'father)
3       (PUTPROP child (+ (GET father 'gg) cost) 'gg)
4       (SET child ff))

1   (DE addtohead (lab ff) ;Called by: orderchildren and reorghead
2       (PROG (list start)
3           (SETQ list head)
4           (UNTIL (NULL list)
5               (COND ( (<= ff (EVAL (CAR list)))
6                       (SETQ head (APPEND start (CONS lab list)))
7                       (RETURN T))
8                   (T (SETQ start (APPEND start (LIST (CAR list)))
9                       list (CDR list)))))
10          (SETQ head (APPEND head (LIST lab)))))

1   (DE reorghead (lab ff) ;Called by: orderchildren
2       (REMOVE lab head)
3       (addtohead lab ff))
```

The definitions specific to A^*, A^{e*} and $A^{\alpha*}$ algorithms do not give rise to programming variations here: the same *basic* program 1.1 is used. As has been stated, the *result* satisfies different properties according to whether evaluations of type A^*, A^{e*} or $A^{\alpha*}$ are employed with this basic program.

Comments for program 1.1
Data structures
Each state will be represented by a *label* and several *property values*. The label will be produced by calling the function GENSYM (at the time when the state is created); hence it will be of type: g_i, e.g. **g238**. There are 4

predefined properties for each state: state, father, gg, ff. For example, Fig. 1.3 below shows all the information associated with a state in the 8-puzzle.

The sets head and store of algorithm 1.3 will now be lists of labels. For instance, at some point head might have the value: (g213 g208 g223). The list head will be kept ordered (later on we shall see how and why) while the order of the list store will not be significant.

General procedures and specific procedures
Program 1.1 is made up of eleven general procedures (from ordsrchA to reorghead). But calls are made to five other procedures whose body will be redefined for each individual application: initialstate, develop, hh, goal and displaystate. Each of the general procedures is now detailed, followed by the specific procedures.

General procedure
ordsrchA
Lines 3 and 4 carry out initialisations. In particular they implement instructions 1 to 5 of algorithm 1.3. ndevels will count the nodes developed; ncreated will count the nodes created; store will take every node which has just been developed (or redeveloped). Let us look at the call of the procedure create: (create (initialstate) NIL 0). The evaluation of the first argument will call the specific procedure initialstate; the latter might supply, for example, the representation of the initial state of the board in the 8-puzzle. The procedure create takes three arguments: the *representation* of a state, the *label* of its father (in this case the empty list: there is no father for the initial state), and the *cost* of the shortest path currently known from the initial state to the state created (in this case 0). The procedure create returns a new *label* of the form g$_i$, such as:g217.

Lines 5 to 10 implement lines 6 to 42 of algorithm 1.1. At line 6, choice receives the result of the call to focus: if a label in head denotes a goal-state, focus returns it, otherwise focus returns the digit 1. In line 7, when the label for a goal-state has been found in head, the final procedure which displays the result is called. Line 8 calls for the label referring to the node which is about to be developed (this is the first labelled node in head, as we shall see) to be transferred from head to store. Examine the expression (develop (GET (CAR store) 'state)) in line 10. develop is a specific procedure; it takes the representation of a state m and returns a (possibly empty) list in which each odd numbered element is the *representation* for a state n which is a child of m, while each even numbered element is the cost associated with the arc (m,n). The orderchildren procedure carries out the task described by instructions 19 to 40 of algorithm 1.3. In line 10 its first argument will be the label of the current developed state, taken from the head of store (store has just been lengthened by the call of transferfather). Line 11 has no equivalent in the original algorithm 1.3.

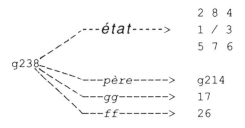

Fig. 1.3.

General procedure **create**
The **create** procedure takes three arguments: the *presentation* of a state, its
father's *label* and the *cost* of the shortest path from the initial state to the
created state currently known. **create** returns a new *label* of the form g_i. This
new label referring to the state created is produced at line 3. Lines 4 to 7
initialise the values of the 4 properties attached to the state: **state, father, gg**
and **ff. ff** is represented by setting the *atomic value* of the label atom directly,
while **state, father** and **gg** are represented by setting three *atomic properties* of the label atom. Lines 5 to 7 implement instructions 36 to 38 of
algorithm 1.3.

General procedure **focus**
If a label in **head** denotes a goal-state, **focus** returns it, otherwise **focus**
returns the digit 1. The labels which make up the **head** list are sorted such
that the values $ff(n_i)$ of states n_i for which they stand are ordered from lowest
to highest. Thus, in line 3, the identifier **ff** takes the smallest value of $ff(n_i)$
(remember that the LISP value of a label atom standing for state n_i is $ff(n_i)$).
Line 5 relies on the fact that the specific procedure **goal?** returns T or ()
depending on whether its argument does or does not represent a goal-state;
the combined work of lines 4 and 5 corresponds to the test present in
instruction 8 of algorithm 1.3. However, there is no point in actually building
the set called **focus** which appears in algorithm 1.3.

General procedure **result**
This is the implementation of instructions 9 to 12 of algorithm 1.3. When
called, the parameter **lab** (label) stands for a goal-state. The cost of the path
found is obtained and displayed by line 2. In line 3 the specific procedure
displaystate is applied to the state representation.

General procedure **transferfather**
This procedure's task is to transfer the currently developed state from **head**
to **store** (instructions 16 and 17 of algorithm 1.3. No parameters are passed
to it since the convention is that the currently developed state will be at the
head of the list called **head**.

General procedure **orderchildren**
This implements instructions 19 to 42 of algorithm 1.3. The father argument
is a state label; childlist is a list in which each odd numbered element is a
child-state (a state *representation* and not a *label*) while each even numbered element is the cost of the arc joining the father-state to the preceding
child-state. Instruction 20 of algorithm 1.3 was nondeterministic; in orderchildren the child-states are considered in the order in which they appear in
childlist.

Lines 5 to 8 carry out instructions 22 to 26. In line 5 exists? tests whether
the first child-state already belongs to head; to be more precise: the first
argument to exists? is the representation of the first child-state, while the
second argument is a list of labels through which the state representations
are accessed; if the test succeeds, exists? returns the label in head which
corresponds to the child-state, otherwise NIL.

In line 6 ffreduced? receives arguments in the following order: the label
of a state (call it n), the label of a state which is the father of that state (call it
m) and the cost of the arc (call it c(m,n)) which links the father to the child; if
it turns out that: gg(m)+c(m,n)<gg(n), then ffreduced? returns a new
value for ff(n), namely: gg(m) +c(m,n)+hh(n), otherwise ffreduced?
returns NIL. Line 7 asks for the values of the properties father, gg and ff
belonging to the state whose label is lab to be updated. Line 8 calls for head
to be reorganised following the assignment of a new ff value to the state
labelled by lab: head must be maintained in increasing order of the
associated ff values; and so lab must be removed from its present position
and sorted back in.

Lines 9 to 12 are the implementation of instructions 27 to 33. In line 10,
exists? tests whether the first child-state already belongs to store. Lines 11
and 12 are identical to 6 and 7. Procedure addtohead, which is called at line
13, performs the same task as reorghead except that it has no need to
extract lab from head before inserting it; note that, unlike instruction 29 in
algorithm 1.3, in practice it is not absolutely necessary to take lab out of
store when it is to be put back into head.

Lines 14 to 17 perform instructions 34 to 39. Line 15 is only evaluated if
the first child-state is not already to be found in either head or store; it is
then necessary to set up all the information (state, father, gg, ff) which
defines a new state; note that when create is called here, the formal
parameter gg is passed the gg value of the father-state incremented by the
cost of the arc joining the father to the child.

General procedure **exists?**
This is given a state representation and a list of labels. It tests whether a
label exists in the list which refers to an identical state. If so, exists? returns
the label found, if not, ().

General procedure **ffreduced?**
ffreduced? receives arguments in the following order: the label for a state n
which already belongs to head or store, the label for a state m which is the

father of n, and the cost of arc c(m,n) which joins father and child. Since
it has appeared previously, n already possesses a value ff(n). If
gg(m)+c(m,n)<gg(n) then ffreduced? returns a new value for ff(n),
namely: gg(m)+c(m,n)+hh(n), otherwise ffreduced? returns (). In line 3,
(EVAL child) returns the old value of ff(n), while (GET child 'gg) returns
the old value of gg(n); the difference is just hh(n); note that in A algorithms
hh(n) depends only on n, unlike gg(n) and ff(n) which also depend on the
state of the algorithm.

General procedure **updatenode**
This implements the 3 groups of instructions: 23 to 25, 30 to 32 and 36 to 38 in
algorithm 1.3. The label received in the parameter child refers to a child-
state, n, of the state, m, referred to by the label received in the parameter
father; cost receives c(m,n); ff receives ff(n).

General procedure **addtohead**
This puts the label symbolised by lab into the head list immediately in front
of the first label referring to a state, n, whose associated value ff(n) is greater
than or equal to that symbolised by ff.

General procedure **reorghead**
Before calling addtohead, this removes the label symbolised by lab from
head.

Specific procedure **initialstate**
Called by ordsrchA. No calling arguments. Must return the agreed representation for the initial state of the particular problem under consideration; for instance, an 8-puzzle board encoded as a list of digits such as: (2 8 3
1 6 4 7 / 5).

Specific procedure **develop**
Called by ordsrchA. Its single argument represents a state to be developed;
for instance, an 8-puzzle board: (2 8 3 1 6 4 7 / 5). The procedure is to return
a list, each odd numbered element of which is a representation of a child-
state while each even numbered element is the cost associated with the arc
joining the father-state to the preceding child-state. If the state to be
developed has no child, the procedure will return (). For example, developing the previous 8-puzzle state and supposing that vertical moves cost 3 while
horizontal moves cost 2, one might get:

((2 8 3 1 6 4 / 7 5) 2 (2 8 3 1 / 4 7 6 5) 3 (2 8 3 1 6 47 5 /) 2)

Specific procedure **hh**
Called by create. Its sole argument represents a state to be developed. It
returns the value of the heuristic function hh for that state.

Specific procedure **goal?**
Called by **focus**. Its sole argument represents a state. It returns T or NIL according as this state is or is not a goal.

Specific procedure **displaystate**
Called by **result**. Its only argument represents a state. It shows the state representations in a manner suitable to the application.

Adapting the program to monotonic heuristic functions
To take advantage of monotonicity (see §1.2.1.1), replace lines 10 to 14 of **orderchildren** by: (NULL (exists? (CAR childlist) store))).

1.4   TRIALS

In this section, program 1.1 will be illustrated by being applied to two simple examples. For each example, the *specific procedures* which are needed to complete the set of *general procedures* will be written, and an execution session will be shown. A variant of program 1.1 will also be proposed which is more suitable for following the details of the heuristic search performed by the basic algorithm.

1.4.1   First example
The *tree* graph in Fig. 1.4 below represents the whole of a state space. The digits stand for the costs associated with the arcs. The node **m** is a goal-state.

1.4.1.1   Specific procedures
The initial state will simply be denoted by: a. Hence:

   (DE **initialstate** () 'a)

The main part of Fig. 1.4 is encoded by the list:

   (a (b 1 c 1 d 1) c (e 1 f 1 g 1) e (h 1 i 1) f (j 1 k 1) j (l 1) l (m 1 n 1))

Hence:

   (DE **develop** (state)
       (CADR (MEMBER state '(a (b 1 c 1 d 1) c (e 1 f 1 g 1) e (h 1 i 1)
                            f (j 1 k 1) j (l 1) l (m 1 n 1)))))

hh is defined by means of a list in which each state precedes the hh value associated with it. Thus:

   (DE **hh** (state)
       (CADR (MEMBER state
              '(a 4 b 5 c 3 d 5 e 3 f 3 g 4 h 3 i 4 j 2 k 4 l 1 m 0 n 2))))

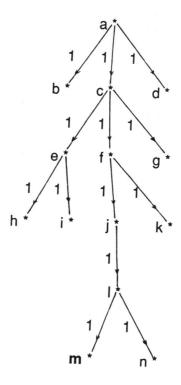

Fig. 1.4 — State space for the first example.

Finally:

(DE **goal?** (state) (EQ state 'm))

(DE **displaystate** (state) (PRINT "node " state))

1.4.1.2   Execution session

The file containing the sets of general and specific procedures is presumed to have been previously loaded. Session 1.1 below starts with the call: (ordsrchA). Once this procedure had executed, the user requested the values of the atoms ndevels and ncreated (which are in effect global variables in program 1.1).

Session 1.1
(ordsrchA)
Here is one solution whose value is 5
node m
obtained from
node l
obtained from
node j

obtained from

node f

obtained from

node c

obtained from

node a

= t

ndevels

= 5

ncreated

= 12

1.4.2   Second example

The *non* arborescent graph of Fig. 1.5 represents the whole of another state space. Nodes **m** and **p** are goal-states.

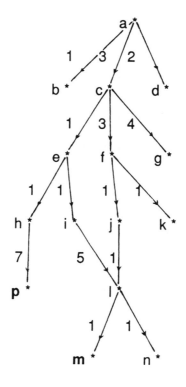

Fig. 1.5 — State space for the second example.

1.4.2.1   *Specific procedures*

We shall reuse:

(DE **initialstate** () 'a)

The main part of Fig. 1.5 is encoded by the list:

(a (b 1 c 3 d 2) c (e 1 f 3 g 4) e (h 1 i 1) f (j 1 k 1) h (p 7) i (l 5) j
   (l 1) l (m 1 n 1))

hence:

(DE **develop** (state)
   (CADR (MEMBER state '(a (b 1 c 3 d 2) c (e 1 f 3 g 4) e (h 1 i 1)
                          f (j 1 k 1) h (p 7) i (l 5) j (l 1)
                          l (m 1 n 1)))))

hh is again defined by means of a list in which each state precedes the hh value which is associated with it. Hence:

(DE hh (state)
   (CADR
      (MEMBER state
              '(a 4 b 5 c 3 d 5 e 3 f 3 g 4 h 3 i 4 j 2 k 4 l 1 m 0 n 2 p 0)))))

Finally:

(DE **goal?** (state) (OR (EQ state 'm) (EQ state 'p)))

(DE **displaystate** (state) (PRINT "node " state))

1.4.2.2 Execution session

The file containing the sets of general and specific procedures is presumed to have been previously loaded. Here is the trace obtained from the call (ordsrchA):

Session 1.2
(ordsrchA)
Here is one solution whose value is 9
node m
obtained from
node l
obtained from
node j
obtained from
node f
obtained from
node c
obtained from
node a
= t
ndevels

= 10
ncreated
= 15

1.4.3   Variant for following the development of the search graph

Program 1.1 gives no details as to the search carried out, other than the final values of ndevels and ncreated. Program 1.2 below includes printing commands so that the evolution of the search graph can be followed.

Program 1.2

```
1    (DE ordsrchA
2        (PROG (ndevels ncreated node head store choice)
3            (SETQ ndevels 0 ncreated 0 store NIL
4                  head (LIST (create (initialstate) NIL 0)))
                                       ;initialstate:specific procedure
            (UNTIL (NULL head)
6                (SETQ choice (focus))
7                (COND ((NOT (NUMBERP choice)) (RETURN (result
                 choice))))
8                (transferfather)
9                (SETQ ndevels (1+ndevels))
10               (TERPRI)
11               (PRIN "development (no. " ndevels
                        ", ff=" (EVAL (CAR store)) ") of node: ")
12               (orderchildren (CAR store)
                             (develop (PRIN (GET (CAR store) 'state)))))
                                  ;develop: specific procedure
13           (PRINT "failure")))

1    (DE create (state father gg)    ;Called by: ordsrchA
2        (PROG (lab)
3            (SETQ ncreated (1+ncreated) lab (GENSYM))
4            (TERPRI)
5            (PRIN "creation (no. " ncreated ") of node " lab "=" state
                  "with value ff=")
6            (PUTPROP lab state 'state)
7            (PUTPROP lab father 'father)
8            (PUTPROP lab gg 'gg)
9            (PRIN (SET lab (+ gg (hh state))))
                     ;hh: specific procedure
10           (RETURN lab)))

1    (DE result (lab)    ;Called by: ordsrchA
2        (PROG ()
3            (TERPRI) (TERPRI)
4            (PRINT "Here is one solution whose value is " (GET lab 'gg))
5            (TERPRI) (displaystate (GET lab 'state))
6            (UNTIL (NULL (SETQ lab (GET lab 'father)))
7                (TERPRI) (PRINT "obtained from ")
8                (TERPRI) (displaystate (GET lab 'state)))))
```

```
1    (DE updatenode (child father cost ff)   ;Called by: orderchildren
2        (TERPRI)
3        (PRIN "update of " child "=" (GET child 'state)
                "(==>ff=" ff ") from " father "=" (GET father 'state))
4        (PUTPROP child father 'father)
5        (PUTPROP child (+ (GET father 'gg) cost) 'gg)
6        (SET child ff))
```

Compared with program 1.1 the modifications only affect procedures ordsrchA, create, result and updatenode; procedures focus, transfer-father, orderchildren, exists?, ffreduced?, addtohead and reorghead are retained in full. The new parts are underlined.

1.4.3.1 *Application to the first example*
The file containing the sets of general procedures (program 1.2) and specific procedures relevant to the first example is presumed to have been loaded prior to the call ordsrchA.

Session 1.3
(ordsrchA)
creation (no. 1) of node g119=a with value ff=4
development (no. 1, ff=4) of node: a
creation (no. 2) of node g120=b with value ff=6
creation (no. 3) of node g121=c with value ff=4
creation (no. 4) of node g122=d with value ff=6
development (no. 2, ff=4) of node: c
creation (no. 5) of node g123=e with value ff=5
creation (no. 6) of node g124=f with value ff=5
creation (no. 7) of node g125=g with value ff=6
development (no. 3, ff=5) of node: f
creation (no. 8) of node g126=j with value ff=5
creation (no. 9) of node g127=k with value ff=7
development (no. 4, ff=5) of node: j
creation (no. 10) of node g128=l with value ff=5
development (no. 5, ff=5) of node: l
creation (no. 11) of node g129=m with value ff=5
creation (no. 12) of node g130=n with value ff=7
Here is one solution whose value is 5

node m

obtained from

node l

obtained from

node j

obtained from

node f

obtained from

node c

obtained from

node a

= t

Recall that label symbols, such as g143, come from calls to GENSYM in procedure **create**. The number given to the first node to be created (in this case **143**) depends on the number of times GENSYM happens to have been used prior to the call of ordsrchA.

Fig. 1.6 shows the final search graph built from the trace of session 1.3.

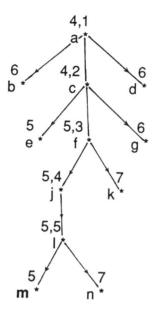

Fig. 1.6 — Final search graph for the first example.

Associated with each node that was created are one or two numbers: the first is the final ff value and the second (if there is one) is the development number. Note that in the general case a node could be developed several times.

Note: when choosing between sibling-nodes e and f, which have the same associated ff value (5), the program gave preference to sibling f which was created last and so placed in front of its sibling in **head** (see **develop** and **orderchildren**, lines 16 and 17). As was stressed in §1.2.1, the basic algorithm 1.3 *does not determine which* state to develop among those which minimise ff. Programs 1.1 and 1.2 *do determine* this state according to a simple technique (which could be reviewed — for a criterion for choosing between states with equal ff values, see [FARRENY & GHALLAB 1986]).

1.4.3.2   Application to the second example

The general procedures of program 1.2 and the procedures specific to the
second example are presumed to have been previously read in.

Session 1.4
(ordsrchA)

creation (no. 1) of node g104=a with value ff=4
development (no. 1, ff=4) of node: a
creation (no. 2) of node g105=b with value ff=6
creation (no. 3) of node g106=c with value ff=6
creation (no. 4) of node g107=d with value ff=7
development (no. 2, ff=6) of node: c
creation (no. 5) of node g108=e with value ff=7
creation (no. 6) of node g109=f with value ff=9
creation (no. 7) of node g110=g with value ff=11
development (no. 3, ff=6) of node: b
development (no. 4, ff=7) of node: e
creation (no. 8) of node g111=h with value ff=8
creation (no. 9) of node g112=i with value ff=7
development (no. 5, ff=7) of node: i
creation (no. 10) of node g113=l with value ff=11
development (no. 6, ff=7) of node: d
development (no. 7, ff=8) of node: h
creation (no. 11) of node g114=p with value ff=12
development (no. 8, ff=9) of node: f
creation (no. 12) of node g115=j with value ff=9
creation (no. 13) of node g116=k with value ff=11
development (no. 9, ff=9) of node: j
update of g113=l (==> ff=9) from g115=j
development (no. 10, ff=9) of node: l
creation (no. 14) of node g117=m with value ff=9
creation (no. 15) of node g118=n with value ff=11

Here is one solution whose value is 9

node m

obtained from

node l

obtained from

node j

obtained from

node f

obtained from

node c

obtained from

node a
= t

Fig. 1.7 represents the final search graph constructed from the trace of

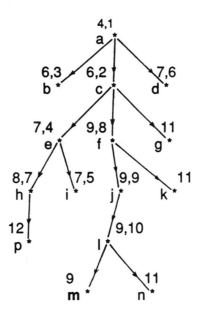

Fig. 1.7 — Final search graph for the second example.

session 1.4. Here too no node is developed more than once. However node l, created as a child of i, is updated to become a child of j.

1.5 A OR A* ALGORITHMS AND THE TRAVELLING SALESMAN PROBLEM

In its canonical form, the *travelling salesman problem* consists of trying to find a minimum cost Hamiltonian circuit of a *complete* graph whose arcs bear costs. This is a path which traverses all of the nodes in a graph *once and once only*, finally returning to the node of departure, such that the sum of the values attached to all of the arcs which make up the path is as small as possible, given that an arc (X,Y) exists between every pair of nodes X,Y in the graph.

Note in passing that in its natural form the travelling salesman problem is better expressed as: find a path which traverses each place in a geographical sector *at least once* before returning to the point of departure. The places can correspond to nodes, the lines of communication to arcs, and the lengths of these lines could be the costs for the arcs. A line of communication does not necessarily exist from one place to another. It can easily be shown that put like this the problem comes back to the preceding canonical form on a graph derived from the graph of places and lines of communication.

One technique for solving the problem in the canonical form above could consist of determining all the Hamiltonian circuits of the graph explicitly (for n nodes: (n-1)! Hamiltonian circuits) and only keeping that or those with the least cost. Implementing such a technique soon becomes impractical as n increases. Note that the number of arcs in the complete graph grows much more slowly: 2(n-1)+2(n-2)+ ... +2 arcs for a complete graph with nnodes, i.e. n(n-1) arcs. It is thus conceivable that we should have the costs of all the arcs yet still be unable to make all the Hamiltonian circuits explicit. Heuristic search techniques can prove useful for limiting the effort required to solve the problem.

Let us recall that the objects which correspond to *arc, path* and *circuit are called edge, chain* and *cycle* in an undirected graph.

In the rest of section 1.5 we shall constantly be looking for a minimum cost Hamiltonian *cycle* in the example of the complete undirected graph G, shown in Fig. 1.8. The actions proposed are directly applicable to any

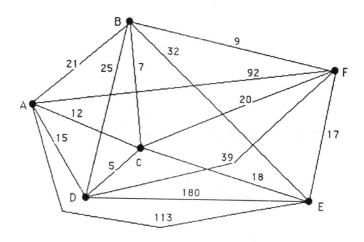

Fig. 1.8 — Example graph G.

undirected graph (each pair of nodes is connected by an edge and not two arcs). Minor alterations would be necessary to deal with directed graphs.

1.5.1   Transition to a problem of state space searching
To find a Hamiltonian cycle in G (Fig. 1.8) one can start from any node, for instance A, then add one of the adjacent nodes to A, for instance B, then add a node adjacent to B and different from A to AB, for instance C, then add a node adjacent to C and different from A and B to ABC, etc...

From this we get the idea of representing an intermediate state in the search for a Hamiltonian cycle by a chain starting from A and only including distinct nodes (the *chain* is called *elementary* in graph theory); that is, a chain of the form A$ where the $ symbol represents a permutation of a (possibly empty) subset of B, C, D, E, F. The initial state of the search will simply be denoted by A.

The goal-states will be of the form A$A where the $ symbol represents a permutation of B, C, D, E, F. Given a state A$ holding less than 6 nodes, its child-states will be of the form A$X where X is any node absent from $. If A$ holds 6 nodes its only child will be A$A. If A$ holds 7 nodes it will not have a child. Thus a state space has been completely defined in which the childless nodes are the goal-states.

Beware: these goal-states correspond only to Hamiltonian cycles. To find a minimum cost Hamiltonian cycle we can envisage attaching values to the arcs in the state space and using an algorithm of type A*. Each arc going from a state A$X to a state A$XY (in the state graph) will naturally have the cost of the arc XY in graph G (the graph of places) as its value. It follows that if the standard term gg(n) of any A* (or indeed A) algorithm is applied to this state space it will measure the cost of the elementary chain encoded by n.

In this representation, the state space associated with a complete graph having n nodes will consist of E(n) nodes, with:

$$E(n) = 1+(n-1)+(n-1)(n-2)+ \dots +(n-1)(n-2)\dots2 + (n-1)! + (n-1)!$$
$$= (n-1)E(n-1)+1.$$

Thus: E(2)=3, E(3)=7, E(4)=22, E(5)=89, E(6)=446 (of which 120 are goal-states) ... E(9)=149921 (of which 40320 are goals).

These numbers underline the fact that, when applying A* algorithms to such state spaces, it is useful to have a "good" evaluation function ff (and hence a "good" function hh) so as to limit the number of states created or developed. A number of heuristic functions hh will now be examined. In each case it will thus be necessary to define a *specific procedure* hh (in the sense of §1.3). On the other hand, the rest of the specific procedures will, except for a few details, still be valid for use with any procedure hh.

1.5.2   Common specific procedures

The problem of finding Hamiltonian chains in a connected undirected graph (whether complete or not) is perfectly defined once a list of edges with associated costs is given.

Let us first initialise a global variable edgecosts for each undirected graph. For graph G of Fig. 1.8 it might be:

```
(SETQ edgecosts '( (a b) 21 (a c) 12 (a d) 15 (a e) 113 (a f) 92
                   (b c) 7 (b d) 25 (b e) 32 (b f) 9
                   (c (d) 5 (c e) 18 (c f) 20
                   (d e) 180 (d f) 39
                   (e f) 17))
```

The procedure findnodes below will be called by:

```
(findnodes edgecosts NIL)
```

and it will set two other global variables, nodelist and nnodes.

```
(DE findnodes (cedges nsofar)
   (PROG (node1 node2)
      (COND ((NULL cedges)
             (SETQ nodelist nsofar nnodes (LENGTH nsofar)))
            (T (SETQ node1 (CAAR cedges) node2 (CADAR cedges))
               (COND ((NOT (MEMQ node1 nsofar))
                      (NEWL nsofar node1)))
               (COND ((NOT (MEMQ node2 nsofar))
                      (NEWL nsofar node2)))
               (findnodes (CDDR cedges) nsofar)))))
```

initialstate will be the procedure which calls findnodes; it will use the first
node encountered (from now on called the initial node) to create and return
the actual initial state. The initial node is saved:

```
(DE initialstate (findnodes edgecosts NIL)
                 (LIST (SETQ initialnode (CAR nodelist))))
```

Obeying the conventions of §1.3, the develop procedure returns a list in
which each child-state is followed by the cost of going from the father to the
child. It produces only one child-state if the father-state has as many nodes
as the graph and if an edge going back to the initial node exists. As agreed
earlier, each state is an elementary chain (we check that a repeated node is
not introduced) coming from the initial node, but the chain is now encoded
by the reversed list of the nodes starting at the initial node. For example, the
list (E C D B A) will correspond to the chain ABCDE. develop calls cost in
order to determine whether an edge exists between two nodes and, if so, to
return the associated cost.

```
(DE develop (state)
   (PROG ()
      (COND ( (EQ (LENGTH state) nnodes)
             (COND ( (NULL (SETQ temp (cost initialnode (CAR state))))
                    (RETURN ())))
             (RETURN (LIST (CONS initialnode state) temp))))
      (APPLY 'APPEND
         (MAPCAR (LAMBDA (add)
                    (COND ((MEMQ add state) ())
                          ((NULL (SETQ temp (cost add
                                  (CAR state)))) ())
                          (T (LIST (CONS add state) temp))))
                 (CDR nodelist)))))

(DE cost (node1 node2)
   (PROG (temp)
      (COND ((OR (SETQ temp (MEMBER (LIST node1 node2) edgecosts))
                 (SETQ temp (MEMBER (LIST node2 node1) edgecosts)))
             (CADR temp)))))
```

If procedure cost is passed nodes B and D as arguments, it looks for the list (B D), or failing that (D B), in edgecosts. If it discovers one or other of these it returns the associated cost, otherwise NIL (so that it can be used on incomplete graphs).

The procedure goal? test whether or not the state holds one node more than the graph:

(DE goal? (state) (EQ (LENGTH state) (+ nnodes 1)))

The displaystate procedure only actually displays the first element of the representative list, since that is the last element to have been added:

(DE displaystate (state) (PRINT "node " (CAR state)))

1.5.3 First example of a heuristic function

Take the state represented by: ABD. To follow this elementary chain until a Hamiltonian cycle is obtained in G (Fig. 1.8) precisely 4 edges are needed (7 nodes to be obtained, less 3 in the state). Now any edge costs at least 5; so the cost of moving from ABD to a goal-state must be greater than or equal to 20. As a general rule:

$\forall n\ hh(n) = $ (number of missing edges) $\times$ (minimum edge cost)

Note: this is a reduced function: $\forall n\ hh(n) \leqslant h(n)$. Hence this function exhibits an A* algorithm: if a Hamiltonian cycle exists (in a complete graph this is certain), the algorithm will find one with a <u>minimum cost.</u>

In dealing with graph G one might be content with:

```
(DE hh (state)
    (* 5 (- 7 (LENGTH state)))))
```

To deal with any graph we can state:

```
(DE hh (state)
    (* minedges (- (+ nnodes 1) (LENGTH state)))))
```

but initialstate and findnodes must be slightly changed as follows (suppose that the cost of edges is always less than 1000):

```
(DE initialstate ()   (findnodes edgecosts NIL 1000)
                      (LIST (SETQ initialnode (CAR nodelist))))
(DE findnodes (cedges nsofar mini)
    (PROG (node1 node2)
        (COND ((NULL cedges)
               (SETQ nodelist nsofar nnodes (LENGTH nsofar)
                     minedges mini))
              (T (SETQ node1 (CAAR cedges) node2 (CADAR cedges))
                 (COND ( (NOT (MEMQ node1 nsofar))
                        (SETQ nsofar (CONS node1 nsofar))))
                 (COND ( (NOT (MEMQ node2 nsofar))
                        (SETQ nsofar (CONS node2 nsofar))))
                 (COND ((< (CADR cedges) mini)
                        (SETQ mini (CADR cedges))))
                 (findnodes (CDDR cedges) nsofar mini)))))
```

The alterations are underlined. Session 1.5 below corresponds to running program 1.1 which has been completed by choosing the above specific procedures for the travelling salesman problem.

Session 1.5
(ordsrchA)

Here is one solution whose value is 85

node f

obtained from

node b

obtained from

node a

obtained from

node d

obtained from

node c

obtained from

node e

obtained from

node f

= t

ndevels

= 72

ncreated

= 138

A Hamiltonian cycle of minimum cost (namely 85) is thus: FBADCEF. As well as calling (ordsrchA), the user asked for ndevels and ncreated. Recall that the state space amounts to 446 nodes of which 326 have children.

1.5.4 Second example of a heuristic function

Consider the elementary chain ABD. In order to complete it at least four edges will be needed whose cumulative cost is certainly greater than or equal to the cumulative cost of the four smallest edges of G. We can thus think of:

$$\forall n \; hh(n) = \text{cumulative cost of the p smallest edges of the graph,}$$
$$\text{if p edges are missing}$$

The A algorithm using this hh function is an A* one again. For graph G the code could be:

```
(DE hh (state) (APPLY '+ (NTHCDR (LENGTH state) '(18 17 15 12 9 7 5))))) or
(DE hh (state) (NTH (− (LENGTH state) 1) '(65 48 33 21 12 5 0)))
```

To deal with any graph one could write:

```
(DE hh (state) (NTH (− (LENGTH state) 1) lreduc))
```

where lreduc is a list which saves the sum of the n smallest edge costs (n=number of nodes in the graph), then the sum of the n−1 smallest costs, then of the n-2 smallest, etc... lreduc is calculated at the beginning of the program: as in the previous example (§1.5.3) initialstate causes the initialisation of nodelist and nnodes, but also lreduc via createlreduc.

```
(DE initialstate ()
      (findnodes edgecosts NIL)
      (createlreduc)
          (LIST (SETQ initialnode (CAR nodelist))))
```

It is the first version of findnodes (§1.5.2) which is used here:

```
(DE findnodes (cedges nsofar)
      (PROG (node1 node2)
          (COND ( (NULL cedges)
                  (SETQ nodelist nsofar nnodes (LENGTH nsofar)))
                (T (SETQ node1 (CAAR cedges) node2 (CADAR cedges))
                  (COND ( (NOT (MEMQ node1 nsofar))
                          (SETQ nsofar (CONS node1 nsofar)))
                  (COND ( (NOT (MEMQ node2 nsofar))
                          (SETQ nsofar (CONS node2 nsofar))))
                  (findnodes (CDDR cedges) nsofar)))))
```

The call to createlreduc leads to calls to lcosts, quicksort, firsts and sums in turn. Let us briefly characterise these procedures by following their effects in the case of graph G; lcosts returns the sub-list of edgecosts composed only of the costs: (21 12 15 113 92 7 25 32 9 5 18 20 180 39 17); quicksort returns: (5 7 9 12 15 17 18 20 21 25 32 39 92 113 180); firsts returns: (17 15 12 9 7 5); sums returns: (65 48 33 21 12 5 0).

```
(DE createlreduc ()
   (SETQ lreduc (sums (firsts nnodes (quicksort (lcosts))))))
```

```
(DE lcosts ()
   (APPLY 'APPEND (MAPCAR (LAMBDA (x)
                          (COND ((NUMBERP x) (LIST x)) (T ())))
                          edgecosts)))
```

To order a list of numbers, quicksort first builds the sub-lists smalls and bigs from those which are less than the head of the list and from those which are greater respectively; it then concatenates the ordered list smalls, the head of the list and the ordered list bigs.

```
(DE quicksort (list)
  (PROG (smalls bigs)
      (COND ((NULL (CDR list)) list)
             (T (smallsandbigs (CAR list) (CDR list))
                (APPEND (quicksort smalls)
                        (CONS (CAR list) (quicksort bigs)))))))))

(DE smallsandbigs (x list)
  (COND ((NULL list) ())
         ((< (CAR list) x) (SETQ smalls (CONS (CAR list) smalls))
                           (smallsandbigs x (CDR list)))
          (T   (SETQ bigs (CONS (CAR list) bigs))
               (smallsandbigs x (CDR list))))))

(DE firsts (n list)
   (COND ((= 0 n) ())
          (T (APPEND (firsts (- n 1) (CDR list)) (LIST (CAR list)))))))

(DE sums (list)
   (COND ((NULL list) '(0))
          (T (CONS (APPLY '+ list) (sums (CDR list)))))))
```

The procedures develop, cost, goal? and displaystate are retained, as well as all the procedures of program 1.1. Here is an execution session:

Session 1.6
(ordsrchA)
Here is one solution whose value is 85
node f
obtained from
node b
obtained from
node a
obtained from
node d
obtained from
node c
obtained from
node e

obtained from

node f

= t

ndevels

= 62

ncreated

= 115

Compare this session with the last one. Naturally, a Hamiltonian cycle with a minimum value of 85 is again found. To be precise, it is the same cycle. The numbers of states developed and created are less than their previous equivalents. However, the second hh function is a little more complex to evaluate than the first.

Could the improvement in ndevels and ncreated have been predicted? Recall the two definitions of hh:

$\forall n\ hh_1(n)$ = number of missing edges x minimum edge cost

$\forall n\ hh_2(n)$ = cumulative cost of the p smallest edges in the graph, if p edges are missing

It is clear that: $\forall n\ hh_1(n) \leqslant 5hh_2(n)$. But, for any graph in general in which a Hamiltonian cycle is sought, it can only be stated that:

$\forall n$, a non-goal state $hh_1(n) < hh_2(n)$

Hence it can not be concluded that hh_2 is *better informed* than hh_1.

1.5.5   Comparison of results of various type A* algorithms

Below is a list of heuristic function definitions which are applicable to the travelling salesman problem (the graph is still presumed to be undirected; the definitions have to be slightly modified for directed graphs), all of which reduce the function h:

Six heuristic functions of type A*

hh_1:   the function equivalent to null (uniform cost)

hh_2:   number of missing edges x minimum edge cost
(this is the function considered in §1.3)

hh_3:   sum of the p shortest edges if p edges are missing
(this is the function considered in §1.4)

hh_4:   sum of the shortest edges coming from the nodes which have not yet been left. For example: 44 for the state ABD (started at A, still have to leave D, C, E and F).

hh_5:   sum of the shortest edges coming from the nodes which have not yet been reached. For example: 49 for the state ABD (still have to reach C, E, F and A)

hh_6:   if n = A\$X, hh(n) = d(X,A): shortest distance between X and A. For example: 28 for the state ABF.

Variants

Several ideas for variants of the above functions immediately spring to mind. For example, one could ensure that the edges involved in calculating hh_3 were different from those already involved in forming the state. In calculating hh_4 and hh_5, one might equally avoid reusing both the edges already involved in the elementary chain represented by the current state and identical edges emanating from 2 nodes to be left or 2 nodes to be reached. Nonetheless it is clear that the evaluation of the heuristic term for these variants becomes more complicated (i.e. more costly in execution time) than in the original functions.

Results

The results obtained by applying program 1.1 (section 1.3), completed by the specific procedures for the heuristic functions hh_1, hh_2, hh_3, hh_4, hh_5 and hh_6 respectively, to graph G (Fig.1.8) are presented below. The specific procedures which are common to all six applications were given in §1.5.2. Graph G is described by **edgecosts**; all six applications use the procedure **findnodes** to determine **nodelist**, **nnodes** and **initialnode**. The implementation of functions hh_2 and hh_3 was described in §1.5.3 and 1.5.4. The Le-Lisp text of the specific procedures for hh_1, hh_4, hh_5 and hh_6 are not given here (the reader will be able to write them without difficulty).

Functions	states developed	states created (ndevels)	algorithm notation (ncreated)
hh_1	88159@1		
hh_2	72	138	@2
hh_3	62	115	@3
hh_4	47	97	@4
hh_5	69	133	@5
hh_6	61	129	@6

Of course, in every case a cycle of minimum cost 85 was obtained. This is always the same cycle **FBADCEF** or **FECDABF** (this has to be so because G has only one minimum cycle).

Monotonic or nonmonotonic functions

Five of the six functions above are monotonic, namely: hh_1, hh_2, hh_4, hh_5 and hh_6. Advantage could be taken of these by means of program 1.1 with the slight simplification suggested at the end of section 1.3.

Algorithms which are or are not better informed

Some of the A^* algorithms obtained with these hh functions can be compared in the sense of the "better informed" notion introduced in §1.2.2.2. In fact, for every non-goal state n (and for any undirected graph of places and lines of communication) it can easily be seen that:

$$hh_1(n) < hh_2 \leqslant hh_3 \leqslant hh_4, \quad hh_1(n) < hh_2 \leqslant hh_3 \leqslant hh_5, \quad hh_1 < hh_6$$

According to the comparison theorem, every state developed by @2, @3, @4, @5 or @6 is developed by @1. The numbers of developed states obtained when applying the algorithms specifically to graph G is in accordance with this result (to verify this, execute algorithms @i with the variant of program 1.2 which displays the details of which developments are performed).

In the general case, @6 cannot be compared with @2, @3, @4 or @5, nor @4 with @5 (remember it is strict inequality that must be respected in the comparison theorem). However, in the particular case of G, one can go a little further. Given the way in which the list of G's edges and associated values happens to have been coded (the identifier edgecosts), the reader will have noticed that node F is the one which is adopted as the initial node for the various elementary paths (i.e. states). Now, the shortest edge coming out of F costs 9 and the minimum of all the edge values is 5; so, if n is a non-goal, $hh_2(n) < hh_5(n)$. Hence any node developed by @5 must be developed by @2; the values of ncreated and ndevels agree with this expectation. In view of the fact that F is the initial node, @4 and @5 are not comparable (even with large inequality); but it would be different if the initial node were C, for the minimum of the edges coming from C is less than the minimum of the edges coming from any other node. If one considers the *state space associated with G which takes C as the initial state*, rather than F, one gets, for a non-goal n: $hh_5(n) < hh_4(n)$, so every state developed by @4 will be developed by @5. In order to retain C as the initial node, we have only to move the first occurrence of C in edgecosts after those of the other nodes. Under these conditions one obtains:

Functions	ndevels	ncreated	algorithms
hh$_4$	30	82	@4
hh$_5$	51	117	@5

These figures conform with what is expected. Here is what is obtained with the other algorithms, in this state space:

Functions	ndevels	ncreated	algorithms
hh$_1$	70	146	@1
hh$_2$	55	125	@2
hh$_3$	49	114	@3
hh$_6$	66	141	@6

Finally, in both the first state space associated with G (initial node: F) and the second (initial node: C), the algorithm which has shown itself to be the most efficient (from the point of view of ncreated and ndevels) of the 6 A* algorithms compared is algorithm @4.

1.5.6   Comparison of results of various type A algorithms

Here are further heuristic functions which are applicable to the travelling salesman problem (still with an undirected graph) which are not less than function h, and thus which lead to A algorithms but not A* ones:

Five other heuristic functions of type A

hh$_7$ if n=A\$X (initial node A), hh(n)=c(X,A) if it is defined, ∞ otherwise. For example: 15 for state ABD.

hh$_8$ number of missing edges x average edge cost

hh$_9$ sum of the average costs of edges coming out of nodes which have not yet been left. For example: 172 for state ABD (D,C,E,F must be exited).

hh$_{10}$ sum of the average costs of edges coming out of nodes which have not yet been reached. For example: 170 for state ABD (D,C,E,F must be reached).

hh$_{11}$ sum of the p largest edges, if p edges are missing.

Results

The table below presents the results gathered from applying program 1.1, completed successively with the specific procedures associated with the five preceding heuristic functions, to graph G (Fig. 1.8). The initial node selected is F: the search is conducted in the first state space previously associated with G (§1.5.5).

Functions	ndevels	ncreated	cycles	costs	algorithms
hh$_7$	49	107	FECDABF	85	@7
hh$_8$	9	22	FEADCBF	166	@8
hh$_9$	14	31	FECDABF	85	@9
hh$_{10}$	6	17	FBADCEF	85	@10
hh$_{11}$	120	209	FBADCEF	85	@11

It is noticeable that 4 out of 5 of these algorithms have found the minimum cycle without any guarantee in advance that they would. To put the result of @8 in context, let us point out the maximal Hamiltonian cycle cost in G is 437. Note the efficiency of @9, and above all of @10!

Monotonic or nonmonotonic functions

The five preceding heuristic functions are nonmonotonic.

1.6   A OR A* ALGORITHMS AND THE 8-PUZZLE

1.6.1   Types of puzzles considered

Many variants of this puzzle exist. From the beginning of this chapter we have used a simple version of the 8-puzzle based on a board of 3×3 squares (each one identical in shape and size), one of which is empty, with the occupied squares being distinguished by the digits 1 to 8. From here onwards, A* algorithms will be applied to puzzles of n×n squares, each identical in shape and size, just one being empty, and the occupied squares all considered to be distinct. An algorithm of type A (but not A*) will be proposed for a board of this type, but having 3×3 squares.

Given an initial configuration and a goal configuration, both of which are completely defined (as in Fig. 1.1 for instance), a sequence of elementary moves will be sought to transform the initial layout into the goal-layout; this sequence may be made up of any moves, but it is to be as short as possible. Such a series will be named a *minimum series*. Fig. 1.2 represents the solution to the problem in Fig. 1.1 (each configuration of the puzzle board denoting one state) by an ordered search method which is not of type A or A*, but which nonetheless produces a minimum series.

To apply A or A* algorithms one must know how to place values on the arcs linking the father-states to their children. For the puzzle problems in which we are interested here (no differences between elementary moves, search for a minimum series), it is natural to attach the same cost to all elementary moves, for example 1.

1.6.2   Transition to a problem of state space searching

A puzzle of n×n squares can be encoded by the concatenation of its lines, that is to say by a permutation of the n×n distinct symbols standing for the squares. For example, the board T:

```
5  0  3
6  2  7
1  8  4
```

where 0 symbolises the empty square) can be encoded by: P1=(5 0 3 6 2 7 1 8 4). P1 will be called a *list of labels*.

If each state is coded by one such permutation, the state space obtained starting from a fixed initial state comes to (n×n)! states at most, giving 362880 states if n=3. The state transformations may be represented by simple edges in an undirected graph since the (permissible elementary) moves are reversible. It can be shown that the graph of all the states which a puzzle of nxn squares can take consists of exactly two connected components, each of which holds (n x n)!/2 states (again giving 181440 if n=3). If the proposed initial and goal states belong to the same connected component a solution will certainly exist, and this solution must be sought in a space of (n×n)!/2 states; if not, no solution exists.

At the start of the chapter it was pointed out that it would be best to work on the moves for the empty square (4 to be taken into account) rather than on those of the occupied squares (4×((n×n)−1) possible moves). We are in fact going to adopt a different representation for the puzzle from that suggested above, one which indicates directly the *position* of the empty square. So for board T:

```
5  0  3
6  2  7
1  8  4
```

we shall use P2=(1 6 4 2 8 0 3 5 7) which is the list of the *positions* occupied by the squares whose labels are, respectively: 0 1 2 3 4 5 6 7 8. The *positions* are numbered:

$$0 \quad 1 \quad 2$$
$$3 \quad 4 \quad 5$$
$$6 \quad 7 \quad 8$$

The *position* of the empty square is given directly by the head of the list P2. Starting from a representation of type P2, the movements of the empty square are simple to program (see procedures right, up, down, left below). P2 will be termed the *list of positions* (an index in this list corresponds to a square label — between 0 and 8 — while each list element is the position — between 0 and 8 — of the square whose label is the element's index).

Programs for the 5 *specific procedures* (see §1.3) initialstate, develop, goal?, displaystate and hh are proposed below. The first 4 are specific to the family of n-square puzzle problems under examination, but independent of the A or A* algorithms applied. The hh procedure will pertain to each A or A* algorithm.

1.6.3   Common specific procedures

To define the problem being dealt with, the global variables initial-board and goal-board will be initialised with the lists of square labels. For example, for the problem defined in Fig. 1.9:

2	8	3		?		1	2	3
1	6	4	-->			8	0	4
7	0	5				7	6	5

Fig. 1.9

the following code will be executed:

```
(SETQ initial-board '(2 8 3 1 6 4 7 0 5)
      goal-board '(1 2 3 8 0 4 7 6 5))
```

The procedure initialstate below assigns the number of squares in the current board to area, the square root of area to size, and (via the procedure transformstate) the list of positions corresponding to the list of labels goal-board to goal-state. In the case of the above example, goal-state will be set to: (4 0 1 2 5 8 7 6 3). Finally, initialstate returns the list of positions corresponding to the list of labels initial-board; for the above example this is: (7 3 0 2 5 8 4 6 1). The variables area, size and goal-state are global.

```
(DE initialstate ()
    (SETQ size (TRUNCATE (SQRT (SETQ area (LENGTH initial-board))))
          goal-state (transformstate area goal-board))
    (transformstate area initial-board))
```

Procedure transformstate below is passed the number of squares in the puzzle, and the list of labels to be transformed (state-before). It constructs a list of positions by finding (via the procedure position) the position (index) of each label in turn in the list of labels state-before. Formally, the roles of the label list and of the position list are interchangeable, and if transformstate is applied again to the result of applying transformstate to a list L, one gets back to the list L; later on displaystate will take advantage of this property.

```
(DE transformstate (number state-before)
    (PROG (state-after)
        (UNTIL (ZEROP number)
          (SETQ state-after
                (CONS (position (1- number) state-before) state-after)
                number (1- number)))
        (RETURN state-after)))

(DE position (number state)
    (PROG (posit)
        (SETQ posit 0)
        (UNTIL (EQ number (CAR state))
            (SETQ posit (1+ posit) state (CDR state)))
        (RETURN posit)))
```

Procedure right receives a list of positions state.

It performs an elementary move of the empty square towards the right, if it is possible to do so. If the empty square is in the right-hand column of the board corresponding to state, right returns (). Otherwise it returns a list whose first element is the position list after the move has been executed and whose second element is always a value of 1 since that is the cost which has been agreed for each move. The call: (subst place0 (1+place0) (CDR state)) attributes the position place0 of the empty square to the label whose position, which is present in the list (CDR state), had the value belonging to (1+ place0) up until now. The procedures left, up and down are analogous to right. The develop procedure returns the concatenation of the results of down, left, up and right (results equal to () are eliminated). Thus, the call: (develop '(7 3 0 2 5 8 4 6 1)) returns: ((6 3 0 2 5 8 4 7 1) 1 (4 3 0 2 5 8 7 6 1) 1 (8 3 0 2 5 7 4 6 1) 1).

```
(DE develop (state)
    (APPEND (down state) (left state) (up state) (right state)))

(DE right (state)
  (PROG (place0)
    (COND ((ZEROP (REM (1+ (SETQ place0 (CAR state)))
                        size))
            (RETURN NIL))
          (T (RETURN
              (LIST (CONS (1+ place0)
                          (SUBST place0 (1+ place0) (CDR state)))
                    1)))))))
```

```
(DE left (state)
   (PROG (place0)
      (COND ((ZEROP (REM (SETQ place0 (CAR state)) size)) (RETURN NIL))
            (T (RETURN
                  (LIST (CONS (1- place0) (SUBST place0 (1- place0)
                                                            (CDR state)))
                     1))))))

(DE up (state)
   (PROG (place0)
      (COND ((< ( SETQ place0 (CAR state)) size) (RETURN NIL))
            (T (RETURN
                  (LIST (CONS (- place0 size) (SUBST place0 (- place0 size)
                                                            (CDR state)))
                     1))))))

(DE down (state)
   (PROG (place0)
      (COND ((>= (SETQ place0 (CAR state)) (- area size)) (RETURN NIL))
            (T (RETURN (LIST (CONS (+ place0 size)
                                   (SUBST place0
                                          (+ place0 size)
                                          (CDR state)))
                        1))))))
```

The procedure goal? compares the position lists representing the current state and the goal-state (which amounts to a comparison between the lists of labels).

```
(DE goal? (state) (EQUAL state goal-state))
```

The displayboard procedure shows the list of $n \times n$ labels with which it is called by displaystate in the form of a board having $n \times n$ squares. displaystate transforms the list of positions state into a list of labels. It takes advantage of the reversibility of the procedure transformstate.

```
(DE displaystate (state)
   (PRINT "board")
   (displayboard (transformstate area state)))

(DE displayboard (list)
   (PROG (lines columns)
      (SETQ lines size columns size)
      (UNTIL (ZEROP lines)
             (PRIN "            ")   ;twelve space characters
             (UNTIL (ZEROP columns)
                    (PRIN (CAR list) " ")   ;1 space character
                    (SETQ list (CDR list) columns (1- columns)))
             (SETQ lines (1- lines) columns size)
             (TERPRI))))
```

1.6.4   First example of a heuristic function: Uniform cost

In order to test the preceding procedures, it only remains to define hh. First of all let us look at the definition:

> (DE **hh** (state) 0)

This characterises a uniform cost algorithm (see §1.2.1.1), which is a special case of an A* algorithm. Session 1.7 below shows the application of general program 1.1 (1.3), completed by the specific procedures which we have just described, to the problem in Fig.1.9.

Session 1.7
(ordsrchA)

Here is one solution whose value is 5

board

```
        1  2  3
        8  0  4
        7  6  5
```

obtained from

board

```
        1  2  3
        0  8  4
        7  6  5
```

obtained from

board

```
        0  2  3
        1  8  4
        7  6  5
```

obtained from

board

```
        2  0  3
        1  8  4
        7  6  5
```

obtained from

board

```
        2  8  3
        1  0  4
        7  6  5
```

obtained from

board

```
        2  8  3
        1  6  4
        7  0  5
```

= t

ndevels
= 33
ncreated
= 61

The solution found is optimal in terms of the number of moves, in accordance with the admissibility theorem.

1.6.5   Second example of a heuristic function: Number of differences
Let us now consider the function hh defined on a state m by:

hh(m) = p   if p occupied squares of state m are not in their
assigned places in the goal-state

It is easy to see that this hh function (which will be termed W hencefor-ward) bears out the characteristic relation of the A^* algorithms. It can be predicted that applying it to the example of Fig. 1.9 will lead to a solution in 5 moves. The function hh=W can be coded simply as:

```
(DE hh (state) (differences (CDR state) (CDR goal-state) 0))

(DE differences (list1 list2 n)
    (COND ((NULL list1) n)
          ( (NEQ (CAR list1) (CAR list2))
            (differences (CDR list1) (CDR list2) (1+ n)))
          (T (differences (CDR list1) (CDR list2) n))))
```

Session 1.8 below shows the application of general program 1.2 (the one which provides a detailed trace, see §1.4.3), completed by the specific procedures of §1.6.3 and by the procedure hh=W, to the problem of Fig. 1.9

Session 1.8
(ordsrchA)
creation (no. 1) of node g104=(7 3 0 2 5 8 4 6 1) with value ff=4
development (no. 1, ff=4) of node: (7 3 0 2 5 8 4 6 1)
creation (no. 2) of node g105=(6 3 0 2 5 8 4 7 1) with value ff=6
creation (no. 3) of node g106=(4 3 0 2 5 8 7 6 1) with value ff=4
creation (no. 4) of node g107=(8 3 0 2 5 7 4 6 1) with value ff=6
development (no. 2, ff=4) of node: (4 3 0 2 5 8 7 6 1)
creation (no. 5) of node g108=(3 4 0 2 5 8 7 6 1) with value ff=5
creation (no. 6) of node g109=(1 3 0 2 5 8 7 6 4) with value ff=5
creation (no. 7) of node g110=(5 3 0 2 4 8 7 6 1) with value ff=6
development (no. 3, ff=5) of node: (1 3 0 2 5 8 7 6 4)
creation (no. 8) of node g111=(0 3 1 2 5 8 7 6 4) with value ff=5

creation (no. 9) of node g112=(2 3 0 1 5 8 7 6 4) with value ff=7
development (no. 4, ff=5) of node: (0 3 1 2 5 8 7 6 4)
creation (no. 10) of node g113=(3 0 1 2 5 8 7 6 4) with value ff=5
development (no. 5, ff=5) of node: (3 0 1 2 5 8 7 6 4)
creation (no. 11) of node g114=(6 0 1 2 5 8 7 3 4) with value ff=7
creation (no. 12) of node g115=(4 0 1 2 5 8 7 6 3) with value ff=5

Here is one solution whose value is 5

board
```
        1  2  3
        8  0  4
        7  6  5
```
obtained from

board
```
        1  2  3
        0  8  4
        7  6  5
```
obtained from

board
```
        0  2  3
        1  8  4
        7  6  5
```
obtained from

board
```
        2  0  3
        1  8  4
        7  6  5
```
obtained from

board
```
        2  8  3
        1  0  4
        7  6  5
```
obtained from

board
```
        2  8  3
        1  6  4
        7  0  5
```
= t

In fact, the solution found not only has the same value but is actually exactly the same as the one obtained with hh being null. We note that the numbers of nodes developed and created (5 and 12 respectively) are less than the equivalent numbers in the case of the null hh (33 and 61); this result was predictable: the second algorithm is manifestly *better informed* than the first in the sense of §1.2.2.2. We also note that the ff values associated with successively developed nodes form a series increasing in size; it can be proved that this is always the case for algorithms with a monotonic function hh; now, W is clearly monotonic, hence the growth observed in the ff values.

1.6.6   Third example of a heuristic function: Sum of distances

The function W does not take proper account of the difficulty involved in causing the *difference* attached to a square to disappear: regardless of whether the square is near to or far from its destination, only one unit of difference is counted. If the destination of a square is x lines and y columns away from its current location, the quantity x+y is the *city-block distance* between the destination and the present position of the square. The hh function (called P from now on) that we are now going to examine adds up the city-block distances associated with each of the occupied squares in the state considered. It is clear that P satisfies the characteristic property of A^*. In addition, for every state m: P(m)<W(m); however, the algorithm associated with P is not better informed than that associated with W (see §1.2.2.2: strict inequality is necessary for the definition). Nonetheless, the algorithm associated with P is better informed than the uniform cost algorithm. Furthermore, P is monotonic.

This function hh=P can be coded by:

```
(DE hh (state) (distances (CDR state) (CDR goal-state) 0))

(DE distances (list1 list 2 n)
    (COND ((NULL list1) n)
          (T (distances (CDR list1)
                        (CDR list2)
                        (+ n (cityblock (CAR list1) (CAR list2)))))))

(DE cityblock (n1 n2)
    (+ (ABS (- (DIV n1 size) (DIV n2 size)))
       (ABS (- (REM n1 size) (REM n2 size)))))
```

By applying the program which is a combination of general program 1.1 (or 1.2), the specific procedures of §1.6.3 and the procedure P to the example of Fig. 1.9, the same solution as with W and the uniform cost is obtained. The numbers of nodes developed and created are 5 and 12 respectively, as with W.

1.6.7   Fourth example of a heuristic function: An A algorithm

Function P does not take proper account of the difficulty involved in exchanging 2 adjoining squares. For a puzzle of 3×3 squares the following hh function defined for any state m can be used:

$$hh(m) = P(m) + 3 \times S(m) \quad \text{where S(m) is evaluated as follows:}$$

— some direction for movement around the central square is chosen;
— with every occupied non-central square in state m, a weight of 0 is associated if it is also non-central in the goal-state and if in both cases the

following square (in the direction of movement chosen) is the same; otherwise a weight of 2 is associated with it;
— if the central square of state m is occupied, a weight of 0 is associated with it if it is already in the place intended in the goal-state, otherwise 1;
— S(m) is the sum of the weights associated with the occupied squares.

For example, if m is the initial-state of the problem given in Fig. 1.9: hh(m)=5+3×9=32. It is clear that this heuristic function does not satisfy the indispensable property of A* algorithms.

The combination of general program 1.1 (or 1.2), the specific procedures of §1.6.3 and the procedure hh=P+3×S gives an A algorithm, so if a solution exists, it will find one. We also observe that, applied to the example of Fig. 1.9, this algorithm discovers the same — optimal — solution as the A* algorithms using null P, W or hh. Furthermore, the numbers of created and developed nodes are 5 and 12 as with P and W.

hh = P + 3 × S can be coded by:

```
(DE hh (state)
     (+ (distances (CDR state) (CDR goal-state) 0)
        (* coeff (+ (internal-zone state) (crown state) ))))

(DE distances (list1 list2 n)
     (COND ((NULL list1) n)
           (T (distances (CDR list1)
                         (CDR list2)
                         (+ n (cityblock (CAR list1) (CAR list2)))))))

(DE cityblock (n1 n2)
     (+ (ABS (- (DIV n1 size) (DIV n2 size)))
        (ABS (- (REM n1 size) (REM n2 size)))))

(SETQ coeff (3)    ;3 if puzzle has 3 × 3 squares, else value to be adjusted
      centre-position (DIV (1- area) 2)
      centre-goal (centre goal-state 0))

(DE internal-zone (state)   ;for puzzle with 2p+1 × 2p+1 squares
     (PROG (center-label)
           (SETQ center-label (label center-position state 0))
           (COND ( (AND (NEQ center-label 0) (NEQ center-label center-goal))
                  (RETURN 1))
                 (T (RETURN 0)))))

(DE label (position state thelabel)
     (COND ((EQ (CAR state) position) thelabel)
           (T (label position (CDR state) (1+ thelabel)))))
```

```
(DE internal? (position)
    (AND (> (DIV position size) 0) (< (DIV position size) (1- size))
         (> (REM position size) 0) (< (REM position size) (1- size))))

(DE crown (state)
    (PROG (limit total templist1 templist2)
        (SETQ limit (1- area) total 0 templist1 (CDR state)
              templist2 (CDR goal-state))
        (UNTIL (ZEROP limit)
            (COND
                ((AND (NOT (internal? (CAR templist1)))
                      (OR (internal? (CAR templist2))
                          (NEQ (label (following-position
                                          (CAR templist1)) state 0)
                ((AND (OR (NEQ (label (following-position (CAR templist2))
                ((AND (OR (NEQ (label goal-state 0))))
            (SETQ total (+ 2 total))))
        (SETQ limit (1- limit) templist1 (CDR templist1)
              templist2 (CDR templist2)))
    (RETURN total)))

(DE following-position (position)
    (COND ( (AND (>= position 0) (< position (1- size)))
          (1+ position))
          ( (AND (EQ (REM position size) 2) (< position (1- area)))
          (+ position size))
          ( (AND (EQ (DIV position size) 2) (> position (- area size)))
          (1- position))
          ( (AND (ZEROP (REM position size)) (> position 0))
          (- position size))))
```

The distances and cityblock procedures are those already introduced in §1.6.6. The function P+3×S is proposed here for a puzzle of 3×3 squares; the coefficient 3 (assigned to coeff) will have to be adjusted for puzzles of other dimensions. The procedure internal-zone returns the weight (1 or 0) assigned to the central square of the board which is represented by the state argument, in accordance with the principle stated above but for boards with (2p+1)x(2p+1) squares. For such boards, the global variables centre-position and centre-goal are respectively set to the number of the central position in the board (numbering from left to right and top to bottom, as in §1.6.2) and to the label of the central square in the goal-board; the initialisation of these variables could be integrated into initialstate. Procedure label returns the label of the square whose position is position in the board represented by state. The internal-zone procedure could be defined differently for puzzles with more than 3×3 squares. Procedure crown returns the weight assigned to the occupied squares on the perimeter of the board represented by the argument state, following the principle laid out above but for boards of any size. internal? returns () if the position is on the outside of the board, or a value different from () if not.

following-position returns the number of the position which comes after the one given by the argument position on the perimeter in the agreed direction of movement (remember that the positions are numbered from left to right and top to bottom).

1.6.8 Some other comparative trials

The table below gives the length of the solution obtained and the number of nodes developed and created by the last 3 of the preceding algorithms (called A1, A2 and A3) when they are applied to the problem represented in Fig. 1.10.

hh functions	States developed (ndevels)	States created (ncreated)	Length of the solution	Notation for the algorithm used
W	1299	2192	18	A1
P	171	200	18	A2
$P + 3 \times S$	41	73	18	A3
$P + 3 \times S1$	72	118	24	A4
$P + 4 \times S$	46	81	18	A5
$P + 3 \times S'$	37	69	18	A6
$P + 3 \times S$	36	66	18	A7

Note the efficiency of A3 compared to A1 in particular.

```
2  1  6          ?          1  2  3
4  0  8         -->         8  0  4
7  5  3                     7  6  5
```

Fig. 1.10

Algorithms A4, A5, and A6 are variants of A3. For A4, the weight of the differences between central squares is ignored in calculating S. For A5, the coefficient 3 for S is replaced by 4. In A6, the order of the state transformation operations in develop is altered; the order: up, right, down, left is substituted for: down, left, up, right. The table shows the results of applying these three algorithms to the same example of Fig. 1.10. On the other hand, algorithm A7, which is identical to A3, is applied to the inverse problem of that in Fig. 1.10. Algorithms A1, A2, A3, A5, and A6 all find the same solution; algorithm A7 finds the inverse solution.

Notes

— A1 was implemented on a Macintosh Plus (one megabyte of memory) with the Le-Lisp dialect. Executing the program (*as described here*) takes several hours.

— In order for A3 to deal with large-scale examples, it is a good idea to refine the initial version of the program given, which is rather simplistic. For example, it might be advantageous to calculate the P and S values associated with state m by revising the equivalent values associated with a father-state of m. The time saved will increase in significance as the size of the puzzle grows.

1.7   DIRECTIONS FOR FURTHER INVESTIGATION

Numerous other heuristic search algorithms are presented in [PEARL 1984] and [FARRENY & GHALLAB 1986]. Here we confine ourselves to offering an insight into quasi-admissible heuristic search algorithms.

In 1982, a variant of A^* algorithms hereafter called A^* was proposed by J. Pearl and J. H. Kim on the one hand, and M. Ghallab on the other, independently of one another. Let us go back to algorithm 1.3 (§1.2.1). The only difference is as follows: at instruction 7, instead of keeping only the elements m such that ff(m) is minimal in focus, all m are kept such that:

$$ff(m) \leqslant (1 + \varepsilon) \min_{m' \in \text{head}} f(m') \quad \text{where } \varepsilon \text{ is a number determined in advance}$$

In particular, the relation between hh and h which is vital to A^* algorithms is also compulsory in A^*_ε algorithms. A^* algorithms are a particular case of A^*_ε where $\varepsilon=0$.

Take a state space which contains at least one goal-state and which satisfies hypotheses 1.1 to 1.3. Let c_0 be the cost of a minimum path from the initial state to the goal-states. It can be proved that any type A^*_ε algorithm applied to this state space terminates by finding a path between the initial state and a goal-state whose cost is less than or equal to $(1+\varepsilon)\ c_0$. This property is called: *quasi-admissibility*.

In $A^*\varepsilon$ algorithms the latitude of choice within the focus set is widened. A secondary criterion (in relation to the ff function) can then be employed to determine the next node to be developed (instruction 15 of algorithm 1.3).

Let us take the example of a program which has to determine an itinerary for a mobile robot. It is probably inopportune to look for a strictly minimum itinerary *in an approximate model of the environment* by calling an A^* algorithm: as soon as the movement space is made discrete (whether this be coarse or fine) and values (whether rough or accurate) or placed on the arcs joining the states, the question is raised as to whether the solutions supplied by A^* are optimal *in relation to the real world*. To reach perfectly identical positions, it is possible that the distances to be covered in the model differ from those in the real world; besides, the distance is not necessarily the only cost factor. It may be that an itinerary found in the approximate model of the environment, whose length is 10% greater than that of the optimal itinerary according to this same model, is in fact optimal in the real world. It may equally be possible for an itinerary that is actually about 10% longer to be more interesting in other respects. In these circumstances, an $A^*\varepsilon$ algorithm could be applied with $\varepsilon=0.1$, and state changes which involve the least direction or height changes by the robot might be chosen.

It is easy to revise program 1.1 (§1.3) so as to make use of $A^*\varepsilon$ algorithms. Three general procedures have to be changed: ordsrchA, focus and transferfather. In the procedure bodies below the modified areas are underlined: the line number is underlined if the line is entirely new. One new procedure, choose, must be defined for each application or for a family of applications. When called (instruction 8 of ordsrchA), the argument to choose is the number of elements in head whose associated ff value is less than or equal to $(1+\varepsilon)$ times the minimum of the ff values associated with all of the elements in head. The number of elements is determined by the procedure focus (instruction 6 of ordsrchA). Recall that the head list is kept ordered such that the associated ff values form an increasing list. By convention, the value to be returned by choose is the index of the element in head chosen for development.

```
1      (DE ordsrchA ()
2          (PROG (node head store choice epsilon)
2.1            (PRINT "Epsilon value ?")
2.2            (SETQ epsilon (READ)
3                    ndevels 0 ncreated 0
4                    head (LIST (create (initialstate) NIL 0)))
5              (UNTIL (NULL head)
6                  (SETQ choice (focus))
7                  (COND ((NOT (NUMBERP choice))
                            (RETURN (result choice))))
8                  (transferfather (choose choice) head NIL)
                    ;choose: new specific procedure to be defined
9                  (SETQ ndevels (1+ ndevels))
10                 (orderchildren (CAR store)
                                  ((develop (GET (CAR store) 'state)))))
11             (PRINT "failure")))


1      (DE focus ()
2          (PROG (list ff count)
3              (SETQ list head ff (* (1+ epsilon) (EVAL (CAR head))) count 0)
4              (UNTIL (OR (NULL list) (> (EVAL (CAR list)) ff))
5                  (COND ((goal? (GET (CAR list) 'state))
                            (RETURN (CAR list))))
6                  (SETQ count (1+ count) list (CDR list)))
7              (RETURN count)))
```

```
1     (DE transferfather (nfather tail head2)
1.1       (COND ((EQ nfather 1) (SETQ store (CONS (CAR tail) store)
1.2                                   head (APPEND head2 (CDR tail))))
1.3              (T (transferfather (1- nfather)
1.4                  (CDR tail)
1.5                  (APPEND head2 (LIST (CAR tail)))))))))
```

When called initially (instruction 8 of **ordsrchA**), **transferfather** is passed the index **nfather**, in the list **head**, of the node to be developed. This node's label is extracted from **head** and put into **store**.

When revised in this way, the program allows a class of algorithms larger than just A_ε^* to be represented since the relation between **hh** and **h** required by A^* algorithms does not enter into the program text. For example, this program can be used to implement $A_\varepsilon^{*\alpha}$ algorithms, that is to say A_ε^* algorithms where the relationship between **hh** and **h** has been relaxed: the only requirement is that: $\forall n$ $hh(n) \leqslant (1+\alpha)h(n)$ as for the $A^{\alpha*}$ algorithms. Take a state space which contains a goal-state and which satisfies hypotheses 1.1 to 1.3. Let c_0 be the cost of a minimum path from the initial state to the goal-states. It may be shown that any type $A_\varepsilon^{\alpha*}$ algorithm applied to this state space will terminate by finding a path linking the initial state and a goal-state whose cost is less than or equal to $(1+\alpha)(1+\varepsilon)c_0$.

2

Expert system generators

Expert systems constitute one of the most conspicuous branches of applied AI. The first section presents a very brief review of the principles of these systems. For basic accounts refer to [WATERMAN & HAYES-ROTH 1978], [BARR & FEIGENBAUM 1982], [HAYES-ROTH 1984], [ALTY & COOMBS 1984], [FARRENY 1985], [BENCHIMOL *et al.* 1986], [WATERMAN 1985]. The second section is given over to the description of several simple *generators* for *diagnostic* type expert systems; the initial version will be progressively extended until a version is obtained which encompasses several basic characteristics of a well-known and especially representative system: MYCIN. In the third section a small *action-planning* type generator is constructed. In the fourth section a *pattern-matching* system is presented as an important module in expert systems which allow variables.

2.1 REVIEW

2.1.1 Expert systems: expert system generators

Expert systems are programs (or perhaps soon machines) which are intended to replace or assist the person in domains where human expertise exists which is (i) insufficiently structured to constitute a precise, certain, and complete method of working that can be transferred directly to computer, and (ii) subject to revision and addition in the light of experience.

Examples of such domains are to be found in the sciences of life (medical diagnosis and treatment, for instance), in the human and social sciences (educational advice, work organisation, justice, . . .), and also in the engineering sciences (creation of manufacturing ranges, conception and conduct of a plan for drilling in prospecting for mines, . . .).

In these domains an important part of the know-how possessed by human experts can be represented in the form of *knowledge units*, often called *rules*. Here is an example of a simplified rule taken from the MYCIN expert system (the first version of which is owed to Edward Shortliffe, USA, 1974) used for the diagnosis and treatment of bacterial illnesses:

> if the site of the culture is blood
> and if the stain of the organism is Gramneg

and if the organism is rod-shaped
and if the patient is a subject at risk
then the probability is 60%
that the organism is pseudomonas aeruginosa

In the (1984 version) MYCIN system around 500 rules were collected. The DART system (IBM, USA, 1981), designed for diagnosing computer breakdowns, encompasses 190 rules. The LITHO system (Ste Schlumberger-France, 1981), intended for analysing geological parameters, includes about 400 rules. The SECOFOR system (Ste Elf-Aquitaine, 1982) is designed to diagnose drilling hitches and has around 250 rules. The TOM system (Ste COGNITECH and INRA, France, 1984), designed for the diagnosis of diseases of the tomato, holds approximately 200 rules.

Each rule is appropriate for a particular class of situations and describes a possible elementary stage in the reasoning of experts in the domain (medicine in the case of MYCIN). Of course, several rules can apply to the same class of situations or to related classes. The collection of rules is not definitively complete or unchanging; on the contrary, it is subject to revisions or adjustments a priori.

Expressions describing specific situations are called *facts*. While rules are pieces of *operating knowledge*, facts are pieces of *assertional knowledge*. Generally speaking, facts condition the applications of rules, serve as input arguments to them, and represent the results of these applications.

A *knowledge expression language* is used to encode the rules and facts. The rules and facts are exploited by the part of the expert system called the *inference engine* (*infer* means produce new information from information already possessed). Broadly speaking, inference engines are kinds of general reasoning machines capable of linking together the results of elementary reasoning particular to an application domain, independently of that domain. The rules are data for the inference engine: they can be corrected, added to or removed, without affecting the engine. This makes it possible to perfect sets of rules which allow the functions of human experts to be reproduced.

The combination of a knowledge expression language and an inference engine is frequently called an *empty system* or a *general system* or an *expert system generator*. New expert systems are built starting from the same generator by totally changing the collection of rules. For example, the generator EMYCIN (Empty MYCIN) combining MYCIN's language and engine was the starting point for building the DART, LITHO, SECOFOR and TOM expert systems mentioned earlier.

2.1.2 Knowledge expression languages, without variables
The most advanced expert systems allow variables to be used at least in the expression of rules; usually, a rule of the form:

$$\text{premise}(x,y \ldots) \rightarrow \text{conclusion}(x,y \ldots)$$

is interpreted as: $\forall x,y \ldots$ IF premise$(x,y \ldots)$ is established THEN conclusion(x,y . . .) is established. For example, in order to make use of the rule: birth = x → age = 1986-x, the variable x will have to take on a value. If the rule is invoked via *forward-chaining*, x will take on the value 1920 (x is said to be *instantiated* by 1920) when the premise is matched with the fact: birth = 1920; a new fact will be constituted: age = 66. The use of variables allows the expression of knowledge to be condensed. In the present chapter, with the exception of the last section concerning *pattern-matching*, only rules (and hence facts) without variables will be used. Here is an example of a rule without a variable:

animal flies and animal lays eggs → animal is bird

In this case, animal flies, animal lays eggs and animal is bird are indivisible propositions: no attempt is made to find and instantiate a variable.

2.2   DIAGNOSTIC-TYPE EXPERT SYSTEM GENERATORS

Many expert systems are developed to execute diagnostic tasks: medical, technical and business diagnosis in particular. In the area of diagnostic expert system generators, the EMYCIN system mentioned above is unquestionably the first software reference. Numerous generators which have now been commercialised were directly inspired by EMYCIN. As examples, take TG1 distributed by the Ste COGNITECH, M1 and S1 distributed by the Ste FRAMENTEC, and Personal Consultant from Texas Instruments. Even substantially reduced versions of EMYCIN have given rise to interesting implementations: one such simple version, implemented in BASIC on a microcomputer, brought about BASIC-PUFF, an expert system linked directly to a spirometer, which is actually used by lung specialists to give a rough interpretation of spirometric examinations (100 rules). The elementary generators presented below progressively integrate several of EMYCIN's principal characteristics.

2.2.1   Outlines of backward-chaining, depth first, and tentative regime engines

Algorithm 2.1 below represents the outline of an inference engine using rules without variables. The premise and conclusion parts of the rules are presumed to be conjunctions of elementary facts, as in the last example.

This engine is started by calling the main procedure retro. The keyword **VARIABLE** precedes declarations of local variables which are in addition to the parameters with which the procedures are called; the keyword **RETURN** forces exit from a procedure, returning the value of the expression which follows the variables fact-base and rule-base are global to all procedures. They are presumed to have been initialised when the engine is started.

This engine invokes the rules in a *backward-chaining* fashion: it cyclically searches for those rules whose conclusion part expresses the fact which is currently to be established (instruction 4 of establish-a-fact), the facts to be established which were put onto the stack most recently are preferred for investigation.

The engine works according to a *tentative regime*: the replacement of a fact to be established by several other facts to be established (instructions 4 to 8 of **investigate-it**) is challenged once one of the latter cannot be established by any rule (dropping into instruction 8 of **investigate-them**).

Algorithm 2.1:
```
 1   PROCEDURE retro
 2      VARIABLES: possibles, a-fact
 3      possibles ‹−− diagnoses
 4      UNTIL possibles = {} REPEAT
 5         a-fact ‹−− a fact chosen from those in possibles
 6         possibles ‹−− possibles minus a-fact
 7         IF establish-a-fact(a-fact) = "success" THEN
 8            WRITE [a-fact "is proved"]
 9            RETURN "success"
10         ENDIF
11      ENDUNTIL
12      WRITE ["no diagnosis obtained"]
```

```
 1   PROCEDURE establish-a-fact(the-fact)
 2      VARIABLE: rules
 3      IF the-fact belongs to fact-base THEN RETURN "success"
 4      rules ‹−− all the rules in rule-base
                  which have the-fact in their conclusion part
 5      RETURN investigate-them(rules)
```

```
 1   PROCEDURE investigate-them(the-rules)
 2      VARIABLE: a-rule
 3      UNTIL the-rules = {} REPEAT
 4         a-rule ‹−− a rule chosen from those in the-rules
 5         the-rules ‹−− the-rules minus a-rule
 6         IF investigate-it(a-rule) = "success" THEN RETURN "success"
 7      ENDUNTIL
 8      RETURN "failure"
```

```
 1   PROCEDURE investigate-it(the-rule)
 2      VARIABLES: facts, a-fact
 3      facts ‹−− all the facts making up the premise of the-rule
 4      UNTIL facts = {} REPEAT
 5         a-fact ‹−− a fact chosen from those in facts
 6         facts ‹−− facts minus a-fact
 7         IF establish-a-fact(a-fact) = "failure" THEN RETURN "failure"
 8      ENDUNTIL
 9      RETURN "success"
```

Note: as a result of instruction 9 of **retro**, the engine stops as soon as a diagnosis is established. This strategy is suitable in the case of mutually exclusive diagnoses. In other circumstances it will be necessary to study the existence of other diagnoses beyond the first one discovered.

Algorithm 2.2

```
 1   PROCEDURE retro
 2     VARIABLES: possibles, a-fact
 3     possibles ‹-- diagnoses
 4     UNTIL possibles = {} REPEAT
 5       a-fact ‹-- a fact chosen from those in possibles
 6       possibles ‹-- possibles minus a-fact
 7       IF establish-a-fact(a-fact) = "success" THEN
 8         WRITE [a-fact "is proved"]
 9         RETURN "success"
10       ENDIF
11     ENDUNTIL
12     WRITE ["no diagnosis obtained"]

 1     PROCEDURE establish-a-fact(the-fact)
 2       VARIABLES: rules, reply
 3       IF the-fact belongs to fact-base THEN RETURN "success"
3.1       IF the-fact belongs to questions THEN RETURN "failure"
 4       rules ‹-- all the rules in rule-base
                   which have the-fact in their conclusion part
4.1       IF rules = {} THEN
4.2         WRITE ["can you confirm that:" the-fact "?"]
4.3         READ [reply]
4.4         IF reply = "no" THEN
4.5           questions ‹-- questions plus the-fact
4.6           RETURN "failure"
4.7         ENDIF
4.8         fact-base ‹-- fact-base plus the-fact
4.9         RETURN "success"
4.10      ENDIF
 5       RETURN investigate-them(rules, the-fact)

1.1   PROCEDURE investigate-them(the-rules)
 2    VARIABLE: a-rule
 3    UNTIL the-rules = {} REPEAT
 4      a-rule ‹-- a rule chosen from those in the-rules
 5      the-rules ‹-- the-rules minus a-rule
 6      IF investigate-it(a-rule) = "success" THEN RETURN "success"
 7    ENDUNTIL
 8    RETURN "failure"

 1    PROCEDURE investigate-it(the-rule, the-fact)
 2      VARIABLES: facts, a-fact
 3      facts ‹-- all the facts making up the premise of the-rule
3.1     IF an element of facts belongs to questions THEN
3.2       RETURN "failure"
 4      UNTIL facts = {} REPEAT
 5        a-fact ‹-- a fact chosen from those in facts
 6        facts ‹-- facts minus a-fact
 7        IF establish-a-fact(a-fact) = "failure" THEN RETURN "failure"
 8      ENDUNTIL
8.1     WRITE [a-rule "allowed" the-fact "to be established"]
```

8.2   fact-base ←−− fact-base plus the-fact
8.3   rule-base ←−− rule-base minus a-rule
9   RETURN "success"

Algorithm 2.2 is obtained by revising the following points of algorithm 2.1 (the instructions corresponding to these revisions are marked by inserted numbers).

Instruction 8.2 of **investigate-it** causes the facts established by applying rules to be stored away; note that when a fact is established, it remains so for ever since the engine never suppresses a fact (for this reason the engine's working is called *monotonic*); repetitions of deductions will be avoided thanks to instruction 8.2 of **investigate-it**. Instruction 8.1 allows the user to follow the deductions made. Instruction 8.3, coupled with 8.2, prevents looping: no rule can be fired more than once. Instructions 4.1 to 4.10 in **establish-a-fact** allow (i) the program user to be interrogated about facts which do not figure in the conclusion of any rule and (ii) use to be made of his affirmative or negative reply. Instruction 4.5 avoids the same question being asked more than once; **questions** is a global variable like **fact-base** and **rule-base**; initially, **questions** will probably be empty; the set of facts which gave rise to negative (i.e. non-positive) replies from the user are stored in it. Note that, if at a given instant the user was not able to confirm that a fact was established, it is presumed that the user's knowledge concerning that fact can never change. Furthermore, if a fact has already been the object of a question it is pointless to try to deduce it by rules, hence instruction 3.1 of **establish-a-fact**. Finally, attempting to establish an element of a rule premise (perhaps via many recursive calls of **establish-a-fact**) is useless once any element of this premise already belongs to questions; hence instructions 3.1 and 3.2 of **investigate-it**.

It is important to clearly understand the circumstances in which the engine will address questions to the user: these questions only ever concern the facts supplied as arguments to **establish-a-fact** which do not appear in the conclusion part of any rule. One can of course think of alternative strategies.

2.2.2   Basic LISP program for an interactive, backward-chaining, depth first, and tentative regime engine

To test out the engine represented by algorithm 2.2, it will be applied to an (imaginary) zoological database suggested in [WINSTON 1981]. Drawing up and supplying the initial knowledge for an application so as to form a specialised expert system may be accomplished by using the language associated with a generator; this is often called *instantiation*.

In the present case, the initialisations of all variables describing the knowledge available at the start (namely: rulebase, factbase, diagnoses, questions) will simply be gathered together in the body of a *instantiation procedure* (**zoolog**) which is to be the main procedure of the expert system:

this is the one that will be called first and that will in turn call retro, the main procedure of the engine.

It has also been decided to give a label to each rule that is introduced (in this example: r1, r2, etc...). The use of labels will help to limit the amount of memory used as a result of manipulating rules: some of the engine's procedures will work on lists of rule labels rather than on rule texts. Later on, the rule labels will serve as external names for interacting with the user. Finally the properties (in the LISP sense) of the label-atoms may serve to store information about each rule (rule used or not, rule active or inhibited, category to which it is related, comments ...).

Instantiation program 2.1

```
(DE zoolog ()
    (SETQ
        rulebase '(r1 r2 r3 r4 r5 r6 r7 r8 r9 r10 r11 r12 r13 r14 r15)
        r1  '( ( (animal has hair)) (animal is mammal))
        r2  '( ( (animal gives milk)) (animal is mammal))
        r3  '( ( (animal has feathers)) (animal is bird))
        r4  '( ( (animal flies) (animal lays eggs)) (animal is bird))
        r5  '( ( (animal eats meat)) (animal is carnivore))
        r6  '( ( (animal has pointed teeth) (animal has claws)
                (animal has forward-pointing eyes)) (animal is carnivore))
        r7  '( ( (animal is mammal) (animal has hooves)) (animal is ungulate))
        r8  '( ( (animal is mammal) (animal ruminates)) (animal is ungulate))
        r9  '( ( (animal is mammal) (animal is carnivore)
                (animal is brown coloured) (animal has dark spots))
                (animal is leopard))
        r10 '( ( (animal is mammal) (animal is carnivore)
                (animal is brown coloured) (animal has black stripes))
                (animal is tiger))
        r11 '( ( (animal is ungulate) (animal has long neck)
                (animal has long legs) (animal has dark spots))
                (animal is giraffe))
        r12 '( ( (animal is ungulate) (animal dark stripes)) (animal is zebra))
        r13 '( ( (animal is bird) (animal does not fly) (animal has long neck)
                (animal has long legs) (animal is black and white))
                (animal is ostrich))
        r14 '( ( (animal is bird) (animal does not fly) (animal swims)
                (animal is black and white)) (animal is penguin))
        r15 '( ( (animal is bird) (animal flies well)) (animal is albatross))
        factbase NIL
        diagnoses
                '( (animal is albatross) (animal is penguin)
                (animal is ostrich) (animal is zebra) (animal is giraffe)
                (animal is tiger) (animal is leopard)))
    (retro))
```

The *engine-program* 2.1 below comes from a direct implementation of algorithm 2.2 of 2.2.1 above.

Engine-program 2.1

```
(DE retro () ;Called by the instantiation procedure. diagnoses is a
    ;global variable initialised in the instantiation program.
    ;questions is global too
        (PROG (possibles afact)
        (SETQ possibles diagnoses questions NIL)
        (UNTIL (NULL possibles)
            (SETQ afact (NEXTL possibles))
            (COND ( (establishafact afact)
                    (TERPRI)
                    (PRINT "*** DIAGNOSIS: " afact)
                    (TERPRI)
                    (RETURN 'success))))
        (TERPRI)
        (PRINT "*** no diagnosis can be obtained")
        (TERPRI)
        (RETURN 'failure)))

(DE establishafact (thefact) ;Called by retro and investigateit.
    ;factbase and questions are global variables (see instantiation program
    ;for the 1st and retro for the 2nd)
    (PROG (rules)
        (COND ((MEMBER thefact factbase) (RETURN T))
              ((MEMBER thefact questions) (RETURN NIL)))
        (SETQ rules (inthen thefact))
        (COND ((NULL rules) (RETURN (ask thefact))))
        (investigatethem rules thefact)))

(DE inthen (thefact) ;Called by establishafact
    ;rulebase is a global variable initialised in the instantiation program
    (PROG (therules result)
        (SETQ therules rulebase)
        (UNTIL (NULL therules)
            (COND ( (MEMBER thefact (CDR (EVAL (CAR therules))))
                    (SETQ result (CONS (CAR therules) result))))
            (SETQ therules (CDR therules)))
        (RETURN result)))

(DE ask (thefact) ;Called by establishafact
    ;factbase and questions are global variables (see the
    instantiation program for the first and retro for the second)
    (TERPRI)
    (PRINT "QUESTION: can you confirm that " thefact " is true ?")
    (COND ( (MEMBER (READ) '(no non niet n))
            (SETQ questions (CONS thefact questions)) NIL)
          (T (SETQ factbase (CONS thefact factbase)) T)))
```

```
(DE investigatethem (therules thefact) ;Called by establishafact
    (PROG ()
        (UNTIL (NULL therules)
            (COND ((investigateit (NEXTL therules) thefact) (RETURN T))))
        (RETURN NIL)))

(DE investigateit (therule thefact) ;Called by investigatethem
    ;factbase and rulebase are global variables
    ;declared in the instantiation program
    (PROG (theconditions afact)
        (SETQ theconditions (CAR (EVAL therule)))
        (COND ((onenotestablished? theconditions) (RETURN NIL)))
        (UNTIL (NULL theconditions)
            (SETQ afact (NEXTL theconditions))
            (COND ((NULL (establishafact afact)) (RETURN NIL))))
        (TERPRI)
        (PRINT "RULE *** " therule " *** DEDUCES THAT: " thefact)
        (SETQ factbase (CONS thefact factbase))
        (NEXTL rulebase)
        (RETURN T)))

(DE onenotestablished? (thefacts) ;Called by investigateit
    ;questions is a global variable initialised in retro
    (PROG ()
        (UNTIL (NULL thefacts)
            (COND ((MEMBER (NEXTL thefacts) questions)
                    (RETURN T))))
        (RETURN NIL)))
```

> The procedure inthen builds up a list of rule labels; the parameter therules of investigatethem takes such a list; the parameter therule of investigateit takes a rule label.

The call to inthen in establishafact implements the *pattern-matching* stage (in backward-chaining) of the inference engine's basic cycle. The set of labels returned by inthen represents the *conflict set*. The *conflict resolution* stage is limited to choosing the first rule of this set. Many engines do the same. Note: the order in which inthen concatenates the labels will affect the way the session goes, in terms of the nature and/or order of the questions asked; however, if the possible diagnoses are mutually exclusive (and, naturally, if the database constituted by the initial rules and facts is coherent), the final result will be identical; it might be interesting here to implement a heuristic of ordering rule labels according to *syntactic* criteria (such as: rule with few facts to test in the premise) or *semantic* criteria (such as: more certain rule). Furthermore, it would be a good idea for the work of inthen to be executed only once, at the time when the rule base is constructed (and when later revisions are made), for each of the facts in the

premises of the available rules (or at least those which can be examined); this
last kind of improvement comes under the topic of *compilation* of databases.
>   As provided for in algorithm 2.2, when a rule is applied it is removed
from rulebase: see the end of investigateit (up to this point it has been
presumed that the rules contained only one elementary conclusion). This
simple deletion means that the possibility of looping later on is avoided, but
it does not allow the user to be informed of the potential loops that have
actually been avoided.
>   The procedure onenotestablished? tests whether one of the facts in
the list argument (rule premise) has previously been the object of a question
to which the user replied in the negative. The investigateit procedure takes
advantage of this eventuality by immediately abandoning the rule under
examination. This behaviour could be extended to all the facts which by one
means (questions) or another (deductions) could not be confirmed.
      Session 2.1 below is obtained after both the instantiation program 2.1
and the engine-program 2.1 have been loaded.

Session 2.1
(zoolog)
QUESTION: can you confirm that (animal flies) is true ?
no
QUESTION: can you confirm that (animal has feathers) is true ?
no
QUESTION: can you confirm that (animal gives milk) is true ?
no
QUESTION: can you confirm that (animal has hair) is true ?
yes
RULE *** r1 *** DEDUCES THAT: (animal is mammal)
QUESTION: can you confirm that (animal ruminates) is true ?
no
QUESTION: can you confirm that (animal has hooves) is true ?
no
QUESTION: can you confirm that (animal has pointed teeth) is true ?
yes
QUESTION: can you confirm that (animal has claws) is true ?
yes
QUESTION: can you confirm that (animal has forward-pointing eyes) is true ?
yes
RULE *** r6 *** DEDUCES THAT: (animal is carnivore)
QUESTION: can you confirm that (animal is brown coloured) is true ?
yes
QUESTION: can you confirm that (animal has black stripes) is true ?
no
QUESTION: can you confirm that (animal has dark spots) is true ?
yes
RULE *** r9 *** DEDUCES THAT: (animal is leopard)
*** DIAGNOSIS: (animal is leopard)
= success

2.2.3　LISP program for an interactive, self-explaining, backward-chaining, depth first and tentative regime engine

Engine-program 2.2 below is development of engine-program 2.1. The alterations to the bodies of already existing procedures are indicated by underlining; the names of new procedures are also underlined.

>　In particular, engine-program 2.1 offers the user facilities for asking various types of questions, either about the database investigated, or about the line of reasoning performed (but for the time being only *outside* of the diagnostic phase). This ability comes under the heading of what is called —a little pompously — the *self-explaining* powers of expert systems.

Certain questions (those posed by calling deducible?, canbeused? or rules?) concern the general knowledge possessed by the expert system, independently of an individual diagnostic consultation; they can be put either *beforehand* or *afterwards*. The question deducible? tells the user whether a given fact can be established by applying the rules, and if so, which ones; the question canbeused? tells the user whether a given fact can be used to apply rules, and if so, which ones; the question rules? obtains the text of rules denoted by labels.

The questions asked by calling how?, why? or use? concern the diagnostic consultation *which has just taken place*. The question how? must be understood as: "how was such-and-such a fact established — if at all?"; the question why? should be taken to mean: "to what end was an attempt made to establish such-and-such a fact — if at all?"; the question use?, depending on whether or not it has any arguments, can mean: "was such-and-such rules applied?" or "which rules were applied?".

A seventh question, facts?, allows access to the state of the fact-base either before or after a consultation session.

>　All of the rules in the "zoological" application have only one conclusion. Besides the example, this hypothesis was exploited by engine-program 2.1 at two levels: (i) when the premise of a rule is satisfied, the only thing established is the fact which led to the invocation of the rule, and (ii) when a rule is applied it is removed from rulebase. On the other hand, engine-program 2.2 will take into account rules containing a conjunction of elementary facts in their conclusion.

>　Engine-program 2.2 will exploit more fully the idea involved in the calls to onenotconfirmed? by remembering all the failures of establishafact (after a question *or* after an attempt at deduction).

To aid the comprehension of engine-program 2.2, we first of all present a consultation session obtained after loading both the engine-program and instantiation program 2.1 above (the user's requests are indicated **here** by the symbol "; this symbol does not appear during the real session).

Session 2.2

```
" (zoolog)
  diagnosis (dia) or general information (gen) ?
" gen
  = ask your questions
```

" (deducible? animal is bird)
            (animal is bird) is deducible by: r4 r3
  = next?
" (deducible? animal is beautiful)
            (animal is beautiful) is not deducible
  = next?
" (canbeusedanimal is beautiful)
            (animal is beautiful) cannot be used to fire a rule
  = next?
" (rules? r13 r99 r11)
  rule: r13
        IF:
        (animal is bird)
        (animal does not fly)
        (animal has long neck)
        (animal has long legs)
        (animal is black and white)
        THEN:
        (animal is ostrich)
        r99 does not denote a rule
  rule: r11
        IF:
        (animal is ungulate)
        (animal has long neck)
        (animal has long legs)
        (animal has dark spots)
        THEN:
        (animal is giraffe)
  = next?
" (zoolog)
  diagnosis (dia) or general information   (gen)
" dia
  QUESTION: can you confirm that (animal flies) is true?
" no
  QUESTION: can you confirm that (animal has feathers) is true ?
" yes
  RULE *** r3 *** DEDUCES THAT: (animal is bird)
  QUESTION: can you confirm that (animal flies well) is true ?
" no
  QUESTION: can you confirm that (animal does not fly) is true ?
" yes
  QUESTION: can you confirm that (animal swims) is true ?
" yes
  QUESTION: can you confirm that (animal is black and white) is true ?
" yes
  RULE *** r14 *** DEDUCES THAT: (animal is penguin)
  *** DIAGNOSIS: (animal is penguin)
  information ? (yes/no)
" yes
  = ask your questions

" (facts?)   (animal is penguin)
            (animal is black and white)
            (animal is swims)
            (animal does not fly)
            (animal is bird)
            (animal has feathers)
    = next?
" (use? r2 r14 r50)
            r2 was not fired
            r14 was fired
            r50 does not exist
    = next?
" (use?)
            rules fired: r14 r3
    = next?
" (how? anibal is bird)
            (anibal is bird) is not established
    = next?
" (how? animal is bird)
            (animal is bird) was deduced from:
            (animal has feathers)
            by applying rule r3
    = next?
" (how? animal has feathers)
            you said so!
    = next?
" (how? animal is penguin)
            (animal is penguin) was deduced from:
            (animal is bird)
            (animal does not fly)
            (animal swims)
            (animal is black and white)
            by applying rule r14
    = next?
" (how? animal is beautiful)
            (animal is beautiful) is not established
    = next?
" (why? animal is carnivore)
            (animal is carnivore) is not established
    = next?
" (why? animal is penguin)
            (animal is penguin) was a diagnosis to be examined
    = next?
" (why? animal has feathers)
            (animal has feathers) was sought in order to prove:
            (animal is bird)
    = next?
" (canbeused? anibal likes elephants)
            (anibal likes elephants) can not be used to fire a rule
    = next?

Engine-program 2.2

```
(DE retro () ;Called by the instantiation program
    (PROG (possibles afact)
        (prologue)
        (TAG abandon
            (UNTIL (NULL possibles)
                (SETQ afact (NEXTL possibles))
                (COND ( (establishafact afact)
                        (writediagno afact)
                        (RETURN (epilogue)))))
            (writenodiagno))
        (epilogue)))

(DE prologue () ;Called by retro
    ;diagnoses, rulesfired, startingfacts, factbase, notestablished and
    ;ruleaims are global. possibles is declared in retro
    (TERPRI)
    (PRINT "diagnosis (dia) or general information (gen) ?")
    (COND ((NEQ (READ) 'dia) (TERPRI) (RETURN "ask your questions")))
    (SETQ possibles diagnoses startingfacts factbase
            rulesfired NIL notestablished NIL ruleaims NIL)
    (MAPCAR (LAMBDA (therule)
            (PUTPROP therule NIL 'thensfired)) rulebase))

(DE writediagno (thefact) ;Called by retro
    (TERPRI) (PRINT "*** DIAGNOSIS: " thefact) (TERPRI))

(DE writenodiagno (thefact) ;Called by retro
    (TERPRI) (PRINT "*** no diagnosis can be obtained") (TERPRI))

(DE epilogue () ;Called by retro
    (TERPRI) (PRINT "information ? (yes/no)")
    (COND ((EQ (READ) 'no) (TERPRI) "*** bye")
            (T "ask your questions")))

(DE establishafact (thefact) ;Called by retro and investigateit
    (PROG (rules result)
        (COND ((acquired? thefact) (RETURN T))
                ((notestablished? thefact) (RETURN NIL))
                ((MEMBER thefact ruleaims)
                    (RETURN (doabandon thefact))))
        (SETQ rules (inthen thefact))
        (COND ((NULL rules) (RETURN (ask thefact))))
        (NEWL ruleaims thefact)
        (SETQ result (investigatethem rules thefact))
        (COND ((NULL result) (NEWL notestablished thefact)))
        (NEXTL ruleaims)
        (RETURN result)))

(DE acquired? (thefact) ;Called by establishafact, how? and why?
    (MEMBER thefact factbase))
```

```
(DE notestablished? (thefact)
    ;Called by establishafact and onenotestablished?
    (MEMBER thefact notestablished))

(DE doabandon (thefact) ;Called by establishafact
    (TERPRI)
    (PRINT "*** ABANDONING ON ENCOUNTERING LOOPING ON: " thefact)
    (EXIT abandon))

(DE inthen (thefact) ;Called by establishafact and deducible?
    (PROG (therules result)
        (SETQ therules rulebase)
        (UNTIL (NULL therules)
            (COND ( (MEMBER thefact (CDR (EVAL (CAR therules))))
                    (NEWL result (CAR therules))))
            (NEXTL therules))
        (RETURN result)))

(DE ask (thefact) ;Called by establishafact
    (TERPRI) (PRINT "QUESTION: can you confirm that " thefact " is true ?")
    (COND ((MEMBER (READ) '(no non niet n))
            (NEWL notestablished thefact) NIL)
        (T (createfact thefact))))

(DE createfact (thefact) ;Called by ask and investigateit
    (NEWL factbase thefact) T)

DE investigatethem (therules thefact) ;Called by establishafact
    (PROG ()
        (UNTIL (NULL therules)
            (COND ((investigateit (NEXTL therules) thefact) (RETURN T))))
        (RETURN NIL)))

(DE investigateit (therule thefact) ;Called by investigatethem
    (PROG (theconditions afact)
        (SETQ theconditions (CAR (EVAL therule)))
        (COND ((onenotestablished? theconditions) (RETURN NIL)))
        (UNTIL (NULL theconditions)
            (NEXTL theconditions afact)
            (COND ((NULL (establishafact afact)) (RETURN NIL))))
        (TERPRI)
        (PRINT "RULE *** " therule " *** DEDUCES THAT: " thefact)
        (createfact thefact)
        (rememberfiring (thenindex thefact (CDR (EVAL therule)) 0) therule)
        (RETURN T)))

(DE thenindex (thefact thenpart index) ;Called by investigateit and
    ;ruleconcludes
    (COND ((NULL thenpart) index)
        ((EQUAL thefact (CAR thenpart)) (1+ index))
        (T (thenindex thefact (CDR thenpart) (1+ index)))))

(DE rememberfiring (index therule) ;Called by investigateit
    (NEWL rulesfired therule)
    (PUTPROP therule (CONS index (GET therule 'thensfired)) 'thensfired))
```

```
(DE onenotestablished? (thefacts) ;Called by investigateit
    (PROG ()
        (UNTIL (NULL thefacts)
            (COND ((notestablished? (NEXTL thefacts)) (RETURN T))))
        (RETURN NIL)))

(DF deducible? !thefact ;Called by the user himself
    (PROG (rules)
        (space)
        (COND ( (SETQ rules (inthen !thefact))
                (PRIN !thefact " is deducible by: ") (writeseries rules)
                (TERPRI))
              (T (PRINT !thefact " is not deducible")))
        'next?))

(DE space () (PRINCH '| | 12))
    ;Called by deducible?, canbeused?, showrule, how?, why? and use?

(DE writeseries (list) ;Called by deducible?, canbeused? and use?
    (UNTIL (NULL list) (PRIN (NEXTL list)) (PRINCH '| |)))

(DF canbeused? !thefact ;Called by the user himself
    (PROG (rules)
        (space)
        (COND ( (SETQ rules (inif !thefact))
                (PRIN !thefact " can be used to fire rules ")
                (writeseries rules)
                (TERPRI))
              (T (PRINT !thefact " cannot be used to fire a rule")))
        'next?))

(DE inif (thefact) ;Called by canbeused?
    (PROG (therules result)
        (SETQ therules rulebase)
        (UNTIL (NULL therules)
            (COND ((if? thefact (CAR therules))
                    (NEWL result (CAR therules))))
            (NEXTL therules))
        (RETURN result)))

(DE if? (thefact therule) ;Called by inif and why?
    (MEMBER thefact (CAR (EVAL therule))))

(DE rules? !labels ;Called by the user himself
    (TERPRI)
    (UNTIL (NULL !labels) (showrule (NEXTL !labels)))
    'next?)

(DE showrule (therule) ;Called by rules?
    (COND ( (MEMBER therule rulebase)
            (TERPRI) (PRINT "rule: " therule)
            (space) (PRINT "IF:") (oneperline (CAR (EVAL therule))) (TERPRI)
            (space) (PRINT "THEN:") (oneperline (CDR (EVAL therule)))
            (TERPRI))
          (T (PRINT therule "does not denote a rule"))))
```

```
(DE oneperline (list) ;Called by showrule and how?
    (MAPCAR (LAMBDA (x) (space) (PRINT x)) list))

(DF how? !thefact ;Called by the user himself
    (PROG (therule)
        (space)
        (COND ((NOT (acquired? !thefact))
                (PRINT !thefact " is not established"))
              ((MEMBER !thefact startingfacts)
               (PRINT !thefact " was established at the start"))
              (T (SETQ therule (ruleconcludes !thefact))
                 (COND ((NULL therule) (PRINT "you said so!"))
                       (T (PRINT !thefact " was deduced from:")
                          (oneperline (CAR (EVAL therule)))
                          (space) (PRINT "by applying rule " therule)))))
        (RETURN 'next?)))

(DE ruleconcludes (thefact) ;Called by how?
    (PROG (rules indices)
        (SETQ rules rulesfired)
        (UNTIL (NULL rules)
            (SETQ indices (GET (CAR rules) 'thensfired))
            (UNTIL (NULL indices)
                (COND ( (EQUAL (NTH (1- (CAR indices))
                                     (CDR (EVAL (CAR rules))))
                          thefact)
                        (RETURN (CAR rules))))
                (NEXTL indices))
            (NEXTL rules))
        (RETURN NIL)))

(DF why? !thefact ;Called by the user himself
    (PROG (rules)
        (space)
        (COND ( (NOT (acquired? !thefact))
                        (PRINT !thefact " is not established"))
              ( (MEMBER !thefact diagnoses)
                (PRINT !thefact " was a diagnosis to be investigated"))
              ( (MEMBER !thefact startingfacts)
                (PRINT !thefact " was established from the outset"))
              (T (SETQ rules rulesfired)
                 (UNTIL (NULL rules)
                     (COND ( (if? !thefact (CAR rules))
                             (SETQ inif T)
                             (PRINT !thefact " was sought in order to
                                             prove:")
                             (writethensfired (CDR (EVAL (CAR rules)))
                                              (CAR rules))))
                     (NEXTL rules))))
        (RETURN 'next?)))
```

```
(DE writethensfired (listthens therule) ;Called by why?
    (MAPCAR (LAMBDA (then) (space) (PRINT (NTH (1- then) listthens)))
        (GET therule 'thensfired)))

(DF use? !therules ;Called by the user himself
    (COND ( (NULL !therules)
            (space) (PRIN "rules fired: ") (writeseries rulesfired) (TERPRI)))
    (UNTIL (NULL !therules)
        (space)
        (COND ( (MEMBER (CAR !therules) rulesfired)
                (PRINT (CAR !therules) " was fired"))
            ( (MEMBER (CAR !therules) rulebase)
                (PRINT (CAR !therules) " was not fired"))
            (T (PRINT (CAR !therules) " does not exist")))
        (NEXTL !therules))
    'next?)

(DE facts? () ;Called by the user himself
    (oneperline factbase) 'next?)
```

Engine-program 2.2 is built as an extension of engine-program 2.1; the main individual points are commented on below.

Interaction in engine 2.2

In this program seven types of questions have been made available to users: deducible?, canbeused?, rules?, facts?, how?, why? and use?. Other, more sophisticated, types could be suggested; one to obtain all the rules mentioning certain combinations of words or one to obtain a complete review in one go of how a fact was proved, etc.

Remember that the questions permitted here have to be asked either before or after the potential diagnostic session, as appropriate, but not *during* it.

Apart from the 7 predefined types of question above, it is possible to ask *crude* question (still before or after a diagnostic session) by entering the name of any of the global variables. For instance, typing the label of a rule is all that is needed to obtain its text, or entering diagnoses will give the list of possible diagnoses.

Engine 2.2 can only ask the user about facts which do not come in the conclusion of any rule (see line 7 of establishafact). This is a *syntactic, blind* or *uninformed* strategy in that it does not take account of the intrinsicnature of the facts but simply of their location in the rule graph. A different strategy would consist of also allowing questions when it turns outthat a fact can not be deduced because, although it is the conclusion of some rules, these rules are not applicable; in this case it would be necessary ask might be called on return from investigateit (line 9 of establishafact). However, such a strategy risks (i) inundating the user with questions and (ii) failing to distinguish between the levels of knowledge (level 0 for knowledgewhich is not the conclusion of any rule, level 1 for that which is theconclusion of at least one rule with level 0 premises, etc...); as an example,if a diagnosis cannot be deduced, this strategy would lead to putting a question about it. It is still a blind strategy. In order to control the appropriateness of questions, some generators provide for strategies which maybe described as *informed*. For example, in EMYCIN or its derivatives each fact belongs in advance to one of the following 4 categories: exclusively askable (useless to look for rules to prove it), exclusively deducible (cannot be asked), to be asked as a matter of priority (do not resort to the rules unless the response does not decide the matter), to be asked only after the attempts at deduction have failed. It is to be noted that for facts belonging to the last 2 categories the following problem arises: would compatible results be obtained by asking the user and by deduction? This problem can be linked to that of the *coherence of knowledge bases.*

System 2.2 only recognises facts which are established (i.e. true) and facts which are not established (i.e. whose truth value is indeterminate), and not facts which are true or false: hence the style of the questions: can you confirm that such-and-such is true?. The engine works by classing everything that is unknown as false. This behaviour belongs to what is termed the *closed-world hypothesis.* Another kind of generator might ask: can you confirmthat such-and-such a fact is false?, treating everything unknown as true.

System 2.2 is purely *propositional*; every fact is an indivisible whole. For this reason, in the absence of a specific rule, the system cannot foresee the connection between the replies to the questions about (bird flies) and (bird does not fly).

When the user cannot reply (that is to say when he cannot — or does not want to — *assert that a fact is true*) the question will not be put again later on. A possible refinement would be to distinguish the case where the user indicates that he definitely cannot answer the question from the case where the user does not rule out the possibility of replying later on, in which case the system would be permitted to ask the question again.

Treatment of multiple conclusions

All of the rules in the "zoological" example contained only a single elementary fact in their conclusion. Engine 2.2, however, is able to interpret rules whose conclusion part is a conjunction of elementary facts (*rules with multiple conclusions*). Let R be a rule containing 2 conclusions, F1 and F2, which is invoked to try and establish F1. If the premise is satisfied, 2 strategies are possible: F1 alone is saved as the current goal or F2 is also saved *in passing*. The second of these strategies introduces forward-chaining stages onto a backward-chaining base: F2 will not be established after searching for a rule which reaches F2 in its conclusion, but as a result of a rule, R, having its premise satisfied independently of any interest in F2. If establishing F2 eventually turns out to be indispensable, the second strategy will have simplified the work that has to be done; on the other hand, if establishing F2 is not vital to reaching the diagnosis, the first strategy will not waste any time or space on it. In engine 2.2 the first strategy was chosen simply to make the transition to engine 2.3 (presented later on) easy; for engine 2.3, which can deal with uncertain knowledge in its reasoning, the first strategy has been adopted because it is distinctly more straightforward to program than the second (see §2.2.5).

From a purely algorithmic point of view, engine 2.1 is just as capable of dealing with multiple conclusion rules as engine 2.2 (invoke the rules through *one* of the conclusions, test the premise of one of them, draw *that* conclusion). But the anti-looping mechanism incorporated in engine 2.1 removes every rule *as soon as one* conclusion has been reached; so this engine cannot fully exploit the deductive power stored up in a rule base containing multiple conclusions.

In engine 2.2, the strategy reserved for handling multiple conclusions is revealed in (i) a new anti-looping mechanism and (ii) a specific organisation for replying to how? questions. These two points are considered in turn.

Anti-looping mechanism
When it applies a rule, engine 2.2 only draws one conclusion: unlike what happens with engine 1.1, the rule is not discarded (though those with a single conclusion could be) in case another of the conclusions has to be established later on. Preventing looping (during execution) is accomplished simply by prohibiting rules from being set off on a fact for which there is an *ongoing* attempt at deduction. Each time that the engine calls up investigatethem, the fact being pursued is stacked on ruleaims: see line 8 of establishafact; each time investigateit is exited ruleaims is popped: see line 11 of establishafact. Line 5 in establishafact looks out for the possibility of a loop: should that be the case, the call to doabandon causes an escape route (named abandon, declared in line 4 of retro, and called at line 4 in doabandon) to be taken.

ruleaims could be written out at that moment; a refinement would be to display also the sequence of rule labels linking the two occurrences of the fact which led to the loop being prevented. Detecting the existence and the terms of a loop (in the course of execution, or even beforehand) also comes from the problem of the *coherence of knowledge bases* (initial knowledge *and* knowledge gained during execution).

Preparing replies to how? questions
In order to keep track of how such-and-such a fact has been established, it is not enough simply to remember the list of rules "fired" (i.e. applied): it is also necessary to note for which conclusion(s) the rule was fired. This has been anticipated in procedure rememberfiring, which is called by investigateit each time a rule is applied, by updating the value (a list) of a property (thensfired) associated with the rule's label; the new value to be assigned is determined by thenindex: this finds the index of the elementary conclusion, fired by the rule, in the body of the "then" part of that rule. To decide how a fact was established, the procedure how? subsequently uses ruleconcludes to consult the value of the thensfired property belonging to only those rules

that were fired; if one of the conclusions marked by this value is the same as the fact being sought, the rule in which it appears is the one that established the fact.

Recording facts definitively established to be false

Engine 2.1 was limited to recording (in questions) those facts which could not be confirmed when submitted to the user. Engine 2.3 in addition records facts which could not be established after investigating the set of rules: the purpose is to avoid attempting to make a deduction about them again. In engine 2.1, using the "zoological" example, the system invokes r15 when trying to establish (animal is albatross); rules r3 and r4 are then tried in order to establish the first condition of r15 (let us suppose vainly so, with a cost of at least 2 questions): (animal is bird); the system thus gives up applying rule r15, then attempts to establish (animal is penguin) by rule r14, which causes a new invocation of rules r3 and r4; thanks to questions, the system discovers that neither of the 2 rules is applicable, without having to ask anything further. In the same circumstances, engine 2.2 does not consider re-invoking r3 and r4 since it first tests whether the fact is already known to be established as false. Such facts are saved in notestablished; this list is updated in establishafact and ask; it is tested by notestablished? and indirectly by establishafact, onenotestablished? and investigateit. This behaviour is based on the (previously mentioned) hypotheses whereby: (i) every fact once established remains so, and (ii) what the user does not know does not change. Since the value *established* or *not established* pertaining to facts does not alter, engine 2.2 is said to have *monotonic* behaviour.

To conclude, we should point out that the value of the thensfired property associated with each rule label is re-initialised, and that some of the lines of engine-program 2.1 have been rewritten in engine 2.2, though their sense remains unchanged.

Engine 2.2 will now be further developed so that it can use uncertain-knowledge, along the lines of the reasoning model proposed originally in MYCIN/EMYCIN.

2.2.4   Uncertain reasoning in the style of EMYCIN

The user of EMYCIN attaches, either explicitly or by default, *a degree of certainty* lying between 1 and −1 to each fact supplied.

The value 1 means that it is absolutely certain that the fact is true, −1 that it is absolutely certain that the fact is false, 0 represents complete uncertainty, and the intermediate value 0.9, for example, would represent the near certainty that the corresponding fact is true.

In addition, the experts who supply the rules for the application associate an *attenuation coefficient* to each one, which lies between 1 and −1 too. The interpretation assigned to this coefficent (AC hereafter) is the following: if the conditions specified by the premise are verified for absolute certain (the premise certain to degree 1) then, *according to the present rule*, the conclusion is certain to degree AC. If the premise is verified to a degree P *greater than a certain threshold* (0.2 in practice), then the conclusion is certain to degree P x AC (again: according to the current rule). In the 1984 version of MYCIN (around 500 rules), the attenuation coefficients are divided up approximately as follows:

$$AC = 1 \qquad 12\%$$
$$0.5 < AC < 1 \qquad 48\%$$
$$0 < AC \leqslant 0.5 \qquad 25\%$$
$$-0.5 \leqslant AC < 0 \qquad 5\%$$
$$-1 < AC < 0.5 \qquad 3.5\%$$
$$AC = -1 \qquad 7\%$$

When a premise is a simple conjunction of elementary facts which can come in a fact base, the premise's degree of certainty, viewed as a whole, is calculated as the minimum of the certainty degrees of the implied elementary facts. For example, take a rule of the form: IF F1 and F2 THEN (0.75) F3; suppose that F1 and F2 are certain to degrees 0.8 and 0.9 respectively; according to this rule, a certainty of 0.6 for F3 is derived.

If one rule has produced a certainty factor CF1 for a fact and if another rule independently produces a certainty factor CF2 for the same fact, then the *cumulative certainty factor* CF associated with the fact after the 2 rules have been investigated is defined by the formulae:

if CF1 and CF2 are positive : $CF = CF1 + CF2 - CF1 \times CF2$
if CF1 and CF2 are negative : $CF = CF1 + CF2 + CF1 \times CF2$

if CF1 and CF2 have opposite signs:
— if $\min(|CF1|, |CF2|) \neq 1 : CF = (CF1 + CF2)/(1-\min(|CF1|, |CF2|))$
— if $\min(|CF1|, |CF2|) = 1 : CF = 1$

This system of formulae is commutative and associative.

Let us complete the previous example: suppose that a second rule asserts: IF F4 and F5 THEN (0.90) F3. After applying the 2 rules, the cumulative certainty factor of F3 is 0.88.

The final cumulative certainty factor of a fact is calculated by trying to explore *all* the rules which include the fact in their conclusion. If *the absolute value* of this final factor is less than a threshold (in practice this is the same threshold of 0.2 as before), the information about the fact is deemed insufficient: it will not be saved away (reducing the final factor to 0 would amount to the same).

Certainty factors are propagated in this manner, from the initial facts (provided by the user during the current session or perhaps saved at the end of previous sessions) all the way to the potential diagnoses (the list of which is possessed by the system).

In the next section a new engine-program is derived from engine 2.2 but now integrates the model of reasoning based on uncertain knowledge which has just been outlined.

We shall continue to consider only premises formed as conjunctions of elementary facts here. Facts which end up (via questions or rule applications) with a negative factor do not participate in the propagation of certainties: because of the first threshold, the application of the rules in whose premise part they appear is forbidden. The certainty degree, negative or not, of every fact, whether or not propagated, is nonetheless of interest, especially when it comes to diagnostic facts. The EMYCIN generator exploits negative factors in a richer and more complex fashion.

It allows premises to be written which are variant functions of the elementary facts implied (allowing not only conjunction but also, for example, variants of negation); this means that a premise can be evaluated to a degree greater than 0.2 though it implies facts with a negative factor.

2.2.5 LISP program for an interactive, self-explaining, backward-chaining and depth first engine, reasoning about uncertain knowledge

Engine 2.3 described below will be tested on a (realistic but limited) knowledge base in mycology. As for engines 2.1 and 2.2, the initialisation of the knowledge base is simply encapsulated in the main procedure, this time called mycolog. Note the position of the attenuation coefficient in the text of the rule and the existence of rules with multiple conclusions. The attenuation coefficients and the certainty factors will here lie in the range of integers $[-100,100]$ instead of in the range of reals $[-1,1]$ which eases porting to microcomputers having LISP dialects that deal only with integers; the coding of the formulae presented above takes account of this change in scale (see the procedures deduce and updatecf).

Instantiation program 2.2

```
(DE mycolog ()
  (SETQ
  rulebase '( r1 r2 r3 r4 r5 r6 r7 r8 r9 r10 r11 r12 r13 r14 r15
             r16 r17 r18 r19 r20 r21 r22 r23 r24 r25 r26 r27)
  r1   '( ((rimose cap) (dull gills) (dull spores)) 33
         (genus inocybe) (genus pluteus) (genus lepiota))
  r2   '( ((bright coloured cap) (no ring) (no volva)
         (stalk joined to cap)) 90 (genus russula) (family russulaceae))
  r3   '( ((bright coloured cap) (ring) (stalk separating from cap)
         (white spore)) 95 (genus amanita) (family amanitaceae))
  r4   '( ((bright coloured cap) (pink spores) (no ring) (no volva)) 70
         (genus pluteus))
  r5   '( ((bright coloured cap) (ochre spores) (cortina)) 50
         (genus cortinarius) (family naucoriaceae))
  r6   '( ((bright coloured cap) (thick gills) (white spores)) 50
         (genus hygrophorus) (family hygrophoraceae))
  r7   '( ((bright coloured cap) (small) (white spores) (thin stalk)
         (filiform stalk)) 90 (genus mycena) (family marasmiaceae))
  r8   '( ((viscid cap) (no stalk) (ochre spores)) 70
         (genus crepidotus) (family tricholomataceae))
  r9   '( ((viscid cap) (no gills) (tubes)) 40
         (genus boletus) (family boletaceae))
```

```
      r10 '( ((tiny) (filiform stalk) (white spores) (flesh resistant)) 70
              (genus marasmius))
      r11 '( ((tiny) (filiform stalk) (white spores) (flesh not resistant)) 70
              (genus collybia))
      r12 '( ((genus rhodotus)) 100 (family rhodophyllaceae))
      r13 '( ((pink spores) (eccentric stalk) (viscid cap)) 100 (genus rhodotus))
      r14 '( ((cupuloform cap) (gelatinous coat on cap) (flesh resistant)) 90
              (genus geopetalum) (family lepiota))
      r15 '( ((separating gills) (ochre spores)) 60
              (genus paxillus) (family boletaceae))
      r16 '( ((separating gills) (pink spores)) 70
              (genus clitopilus) (family tricholomataceae))
      r17 '( ((pliciform gills)) 50 (genus cantharellus) (family cantharellaceae))
      r18 '( ((dull spores) (thick gills)) 50
              (genus gomphidius) (family boletaceae))
      r19 '( ((separating gills) (pink spores)) 40
              (genus leucopaxillus) (family tricholomataceae))
      r20 '( ((black spores) (cap not exposed)) 40
              (genus panaeolus) (family naucoriaceae))
      r21 '( ((pliciform gills) (bulky cap) (obconical cap)) 60
              (genus gomphus))
      r22 '( ((granular flesh) (thick stalk) (fleshy)) 70
              (genus russula) (family russulaceae))
      r23 '( ((genus pluteus)) 100 (family volvariaceae))
      r24 '( ((genus inocybe)) 100 (family naucoriaceae))
      r25 '( ((genus lepiota)) 100 (family agaricaceae))
      r26 '( ((genus collybia)) 100 (family marasmiaceae))
      r27 '( ((genus marasmius)) 100 (family marasmiaceae))
      factbase NIL
      diagnoses '( (family cantharellaceae) (family boletaceae)
                  (family hygrophoraceae) (family tricholomataceae)
                  (family marasmiaceae) (family naucoriaceae)
                  (family amanitaceae)
                  (family volvariaceae) (family lepiota) (family russulaceae)))
  (retro))
```

Below is a consultation session obtained after joint loading of the preceding instantiation program 2.2 and of engine-program 2.3 which is presented afterwards.

Session 2.3

```
(mycolog)

diagnosis (dia) or general information (gen) ?
gen
= ask your questions
(canbeused? white spores)
            (white spores)
            can be used to fire rules r11 r10 r7 r6 r3
= next?
(canbeused? granular flesh)
            (granular flesh)
```

```
                  can be used to fire rule r22
= next?
(deducible? genus marasmius)
              (genus marasmius) is deducible by: r10
= next?
(rules? r9 r29)

rule: r9
          IF:
          (viscid cap)
          (no gills)
          (tubes)

          THEN (40):
          (genus boletus)
          (family boletaceae)

r29 does not denote a rule
= next?
(mycolog)
```

diagnosis (dia) or general information (gen) ?
dia

QUESTION 1: certainty for (pliciform gills) ? (0 if unknown)
80

PROVISIONAL CONCLUSION, AFTER APPLYING RULE: r17
the cumulative certainty factor of (family cantharellaceae) is 40

*** DIAGNOSIS: (family cantharellaceae) is certain to degree 40

QUESTION 2: certainty for (dull spores) ? (0 if unknown)
90

QUESTION 3: certainty for (thick gills) ? (0 if unknown)
60

PROVISIONAL CONCLUSION, AFTER APPLYING RULE: r18
the cumulative certainty factor of (family oletaceae) is 30

QUESTION 4: certainty for (separating gills) ? (0 if unknown)
25

QUESTION 5: certainty for (ochre spores) ? (0 if unknown)
0

QUESTION 6: certainty for (viscid cap) ? (0 if unknown)
40

QUESTION 7: certainty for (no gills) ? (0 if unknown)
70

QUESTION 8: certainty for (tubes) ? (0 if unknown)
50

PROVISIONAL CONCLUSION, AFTER APPLYING RULE: r9
the cumulative certainty factor of (family boletaceae) is 42

*** DIAGNOSIS: (family boletaceae) is certain to degree 42

QUESTION 9: certainty for (bright coloured cap) ? (0 if unknown)
−100

QUESTION 10: certainty for (pink spores) ? (0 if unknown)
−100

QUESTION 11: certainty for (tiny) ? (0 if unknown)
−80

QUESTION 12: certainty for (rimose cap) ? (0 if unknown)
−50

QUESTION 13: certainty for (black spores) ? (0 if unknown)
−90

QUESTION 14: certainty for (cupuloform cap) ? (0 if unknown)
30

QUESTION 15: certainty for (gelatinous coat on cap) ? (0 if unknown)
−80

QUESTION 16: certainty for (granular flesh) ? (0 if unknown)
−100

*** END OF THE DIAGNOSTIC SESSION
information ? (yes/no)
yes
= ask your questions
           (facts?)
           (granular flesh) certain to degree: −100
           (gelatinous coat on cap) certain to degree: −80
           (cupuloform cap) certain to degree: 30
           (black spores) certain to degree: −90
           (rimose cap) certain to degree: −50
           (tiny) certain to degree: −80
           (no stalk) certain to degree: −100
           (pink spores) certain to degree: −100
           (bright coloured cap) certain to degree: −100
           (tubes) certain to degree: 50
           (no gills) certain to degree: 70
           (viscid cap) certain to degree: 40
           (ochre spores) certain to degree: −50
           (separating gills) certain to degree: 25
           (family boletaceae) certain to degree: 42
           (thick gills) certain to degree: 60
           (dull spores) certain to degree: 90
           (family cantharellaceae) certain to degree: 40
           (pliciform gills) certain to degree: 80
= next?
(use? r18 r17 r2)
      r18 was fired
      r17 was fired
      r2 was not fired
= next?
(use?)
      rules fired: r9 r18 r17
= next?
(why? pliciform gills)
      (pliciform gills)
      was investigated to evaluate the certainty of:
      (family cantharellaceae)
= next?
(why? family cantharellaceae)
      (family cantharellaceae) was a diagnosis to be investigated
= next?

(how? family boletaceae)
> I used rule r18
> to conclude that: (family boletaceae).
> This gave a cumulative certainty factor of: 30.
> The question previously asked was 3.
> I used rule r9
> to conclude that: (family boletaceae).
> This gave a cumulative certainty factor of: 42.
> The question previously asked was 8.

= next?
(how? * family boletaceae)
> I used rule r18
> to conclude that: (family boletaceae).
> This gave a cumulative certainty factor of: 30.
> since the rule's premise has 2 elements
> with respective certainty degrees 90 60
> and the attenuation coefficient is 50.
> The question previously asked was 3.
> I used rule r9
> to conclude that: (family boletaceae).
> This gave a cumulative certainty factor of: 42.
> since the rule's premise has 3 elements
> with respective certainty degrees 40 70 50
> and the attenuation coefficient is 40.
> The question previously asked was 8.

= next?

The replies to questions of type **canbeused?, deducible?, why?** and **use?** are similar to those obtained under engine 2.2. Note: in response to **why?** the program supplies only the conclusion(s) which caused the rule to be applied.The responses to **rules?, facts?** and **how?** questions take account of the existence of attenuation coefficients and certainty factors. Two forms of **how?** questions are permitted; to issue the second kind a * must precede the text of the fact, and the reply furnishes greater detail.

In engine-program 2.3 below those parts or procedure names which are new in comparison with engine-program 2.2 have been underlined.

Engine-program 2.3
```
(DE retro () ;Called by the instantiation program
    (PROG (possibles afact cf result)
        (prologue)
        (TAG abandon
            (UNTIL (NULL possibles)
                (SETQ afact (NEXTL possibles) cf (establishafact afact))
                (COND ((determined? cf) (writediagno afact)
                                        (SETQ result T))))
            (COND ((NULL result) (writenodiagno))))
        (epilogue)))
```

```
(DE prologue () ;Called by retro
    ;diagnoses, startingfacts, factbase, rulesfired, notestablished, ruleaims,
    ;nquestion, descendantrules and threshold are global variables.
    ;possibles is declared in retro
    (TERPRI) (PRINT −diagnosis (dia) or general information (gen) ?−)
    (COND ((NEQ (READ) †dia) (TERPRI) (RETURN −ask your questions−)))
    (SETQ possibles diagnoses startingfacts (flatfacts factbase NIL)
          rulesfired NIL notestablished NIL ruleaims NIL
          nquestion 0 descendantrules NIL threshold 20))

(DE flatfacts (thefacts flat) ;Called by prologue
    (COND ((NULL thefacts) flat)
          (T (flatfacts
                 (CDDR thefacts)
                 (CONS (CAR thefacts) (CONS (GET (CADR thefacts 'cf)
                 flat))))))))

(DE determined? (cf) ;Called by retro and ask
    (AND cf (>= (ABS cf) threshold)))

(DE writediagno (thefact) ;Called by retro
    (TERPRI) (PRINT "*** DIAGNOSIS: " thefact "is certain to degree " cf)
    (TERPRI))

(DE writenodiagno (thefact) ;Called by retro
    (TERPRI)
    (PRINT "*** no diagnosis can be obtained"
           " with an absolute certainty factor greater than " threshold)
    (TERPRI))

(DE epilogue () ;Called by retro
    (TERPRI) (PRINT "*** END OF THE DIAGNOSTIC SESSION")
    (PRINT "information ? (yes/no)")
    (COND ((EQ (READ) 'no) (TERPRI) "*** bye")
          (T "ask your questions")))

(DE notestablished? (thefact) ;Called by establishafact and
onenotestablished?
    (MEMBER thefact notestablished))

(DE establishafact (thefact) ;Called by retro and investigateit
    (PROG (rules cf)
        (COND ((SETQ cf (cfacquired? thefact)) (RETURN cf))
              ((notestablished? thefact) (RETURN NIL))
              ((MEMBER thefact ruleaims)
               (RETURN (doabandon thefact))))
        (SETQ rules (inthen thefact))
        (COND ((NULL rules) (RETURN (ask thefact))))
        (NEWL ruleaims thefact)
        (SETQ cf (investigatethem rules thefact))
        (COND ((NULL cf) (NEWL notestablished thefact)))
        (NEXTL ruleaims)
        (RETURN cf)))

(DE cfacquired? (thefact) ;Called by establishafact, deduce,
                          ;detailsfiring, why?
    (PROG (temp)
        (COND ((SETQ temp (MEMBER thefact factbase))
               (GET (CADR temp)
        'cf)))))
```

```
(DE doabandon (thefact) ;Called by establishafact
    (TERPRI) (PRINT "*** ABANDONING ON ENCOUNTERING LOOPING ON:
    " thefact)
    (PRINT "rule stack: ") (oneperline descendantrules)
    (PRINT "fact stack: ") (oneperline (NEWL ruleaims thefact))
                     (EXIT abandon))

(DE inthen (thefact) ;Called by establishafact and deducible?
    (PROG (therules result)
        (SETQ therules rulebase)
        (UNTIL (NULL therules)
            (COND ( (MEMBER thefact (CDDR (EVAL (CAR therules))))
                    (NEWL result (CAR therules))))
            (NEXTL therules))
        (RETURN result)))

(DE ask (thefact) ;Called by establishafact
    (PROG (cfgiven)
        (SETQ nquestion (1+ nquestion))
        (TERPRI) (PRINT "QUESTION " nquestion ": certainty for " thefact
                        " ? (0 if unknown)")
        (SETQ cfgiven (READ))
        (COND ((determined? cfgiven) (createfact thefact cfgiven NIL)
            (RETURN cfgiven)))
        (NEWL notestablished thefact)
        (RETURN NIL)))

(DE createfact (thefact cf flag)
    ;Called by ask and deduce depending on whether flag is NIL or T
    (PROG (label)
        (SETQ label (GENSYM)
              factbase (CONS thefact (CONS label factbase)))
        (PUTPROP label cf 'cf)
        (COND (flag (howrule label cf))
              (T (PUTPROP label (LIST 'question nquestion) 'how)))))

(DE howrule (label cf) ;Called by createfact and update fact
    (PUTPROP label
                (CONS (LIST (CAR descendantrules) cf nquestion)
                      (GET label 'how))
             'how))

(DE investigatethem (therules thefact) ;Called by establishafact
    (PROG (temp)
        (UNTIL (NULL therules)
            (NEWL descendantrules (CAR therules))
            (SETQ temp (investigateit (NEXTL therules) thefact))
            (NEXTL descendantrules)
            (COND ((AND temp (= 100 (ABS temp))) (RETURN temp))))
        (RETURN temp)))
```

```
(DE investigateit (therule thefact) ;Called by investigatethem
    (PROG (theconditions afact cfpremise cf)
        (SETQ theconditions (CAR (EVAL therule)) cfpremise 0)
        (COND ((onenotestablished? theconditions) (RETURN NIL)))
        (UNTIL (NULL theconditions)
            (NEXTL theconditions afact)
            (SETQ cf (establishafact afact))
            (COND ((OR (NULL cf) (< cf threshold)) (RETURN NIL))
                  ((OR (= 0 cfpremise) (< cf cfpremise)) (SETQ cfpremise cf))))
        (deduce therule thefact (CADR (EVAL therule)) cfpremise)))

(DE onenotestablished? (thefacts) ;Called by investigateit
    (PROG ()
        (UNTIL (NULL thefacts)
            (COND ((notestablished? (NEXTL thefacts)) (RETURN T))))
        (RETURN NIL)))

(DE deduce (therule thefact ac cfpremise) ;Called by investigateit
    (PROG (newcf oldcf)
        (SETQ newcf (DIV (* ac cfpremise) 100) oldcf (cfacquired? thefact))
        (COND (oldcf (SETQ newcf (updatecf newcf oldcf))
                     (updatefact thefact newcf))
              (T (createfact thefact newcf T)))
        (message therule newcf thefact)
        (rememberfiring (thenindex thefact (CDDR (EVAL therule)) 0) therule)
        (RETURN newcf)))

(DE updatecf (newcf oldcf) ;Called by deduce
    (COND ( (> (* newcf oldcf) 0)
            (+ newcf
                (COND ((> newcf 0) (- oldcf (DIV (* newcf oldcf) 100)))
                      (T (+ oldcf (DIV (* newcf oldcf) 100))))))
          ((= 100 newcf) 100)
          (T (DIV (+ newcf oldcf)
                  (- 100 (MIN (ABS newcf) (ABS oldcf)))))))

(DE updatefact (thefact cf) ;Called by deduce
    (PROG (label)
        (SETQ label (CADR (MEMBER thefact factbase)))
        (PUTPROP label cf 'cf)
        (howrule label cf)))

(DE message (name cf thefact) ;Called by deduce
    (TERPRI)
    (PRINT "PROVISIONAL CONCLUSION, AFTER APPLYING RULE: " name)
    (PRINT "the cumulative certainty factor of " thefact " is " cf))

(DE thenindex (thefact thenpart index) ;Called by deduce
    (COND ((NULL thenpart) index)
          ((EQUAL thefact (CAR thenpart)) (1+ index))
          (T (thenindex thefact (CDR thenpart) (1+ index)))))

(DE rememberfiring (index therule) ;Called by deduce
    (NEWL rulesfired therule)
    (PUTPROP therule (CONS index (GET therule 'thensfired)) 'thensfired))
```

```
(DF deducible? !thefact ;Called by the user himself
    (PROG (rules)
         (space)
         (COND ( (SETQ rules (inthen !thefact))
                 (PRIN !thefact " is deducible by: ") (writeseries rules)
                 (TERPRI))
               (T (PRINT !thefact " is not deducible")))
         'next?))

(DE space () (PRINCH '| | 12))
    ;Called by deducible?, canbeused?, showrule, how?, writejustif,
    ;detailsfiring, writethensfired, facts?, writefacts, why? and use?

(DE writeseries (list) ;Called by deducible?, canbeused? and use?
    (UNTIL (NULL list) (PRIN (NEXTL list)) (PRINCH '| |)))

(DF canbeused? !thefact ;Called by the user himself
    (PROG (rules n)
         (space)
         (COND ( (SETQ rules (inif !thefact) n (LENGTH rules))
                 (PRINT !thefact) (space)
                 (PRIN " can be used to fire rules "
                      (COND ((> n 1) "the rules ") (T "the rule ")))
                 (writeseries rules)
                 (TERPRI))
               (T (PRINT !thefact " cannot be used to fire a rule")))
         'next?))

(DE inif (thefact) ;Called by canbeused?
    (PROG (therules result)
         (SETQ therules rulebase)
         (UNTIL (NULL therules)
             (COND ((if? thefact (CAR therules))
                    (SETQ result (CONS (CAR therules) result))))
             (SETQ therules (CDR therules)))
         (RETURN result)))

(DE if? (thefact therule) ;Called by inif and why?
    (MEMBER thefact (CAR (EVAL therule))))

(DF rules? !labels ;Called by the user himself
    (TERPRI)
    (UNTIL (NULL !labels) (showrule (NEXTL !labels)))
    'next?)

(DE showrule (therule) ;Called by rules?
    (COND ( (MEMBER therule rulebase)
            (TERPRI) (PRINT "rule: " therule)
            (space) (PRINT "IF:") (oneperline (CAR (EVAL therule))) (TERPRI)
            (space) (PRINT "THEN (" (CADR (EVAL therule)) "):")
            (oneperline (CDDR (EVAL therule))) (TERPRI))
          (T (PRINT therule "does not denote a rule"))))

(DE oneperline (list) ;Called by showrule
    (MAPCAR (LAMBDA (x) (space) (PRINT x)) list))
```

```
(DF how? !thefact ;Called by the user himself
    (PROG (details justification cf)
        (COND ((EQ (CAR !thefact) '*) (SETQ details T) (NEXTL !thefact)))
        (TERPRI) (space)
        (COND ( (NOT (cfacquired? !thefact))
                (PRINT !thefact " is not recorded"))
              ( (EQ (SETQ cf (startingcf? !thefact)) (cfacquired? !thefact))
                (PRINT !thefact " was certain at the start to degree: " cf))
              (T (SETQ justification
                        (GET (CADR (MEMBER !thefact factbase)) 'how))
                  (COND
                        ( (EQ (CAR justification) 'question)
                          (PRINT "you supplied the degree: "
                          (cfacquired? !thefact))
                          (space)
                          (PRINT "in response to the question: "
                                (CADR justification)))
                        (T (writejustif !thefact (REVERSE justification) details)))))
        (RETURN 'next?)))

(DE startingcf (thefact) ;Called by how?
    (PROG (temp)
        (COND ((SETQ temp (MEMBER thefact startingfacts)) (CADR temp)))))

(DE writejustif (thefact justification details) ;Called by how?
    (UNTIL (NULL justification)
        (PRINT "I used rule " (CAAR justification)) (space)
        (PRINT "to conclude that: " thefact ".") (space)
        (PRINT "This gave a cumulative certainty factor of: "
                (CADAR justification) ".")
        (space)
        (COND (details (detailsfiring (CAAR justification))))
        (PRINT "The question previously asked was "
                (CADDAR justification) ".")
        (TERPRI) (space)
        (NEXTL justification))
    (TERPRI))

(DE detailsfiring (therule) ;Called by writejustif
    (PROG (n)
        (PRINT "since the rule's premise has "
                (SETQ n (LENGTH (CAR (EVAL therule))))
                " element" (COND ((> n 1) "s") (T '| |))) (space)
        (COND ((> n 1) (PRIN "with respective certainty degrees "))
              (T (PRIN "with certainty degree ")))
        (MAPCAR (LAMBDA (afact) (PRIN (cfacquired? afact)) (PRINCH '| |))
                (CAR (EVAL therule)))
        (TERPRI) (space)
        (PRINT "and the attenuation coefficient is " (CADR (EVAL therule)) ".")
        (space)))
```

```
(DF why? !thefact ;Called by the user himself
    (PROG (rules)
        (space)
        (COND ( (NOT (cfacquired? !thefact))
                 (PRINT !thefact " is not recorded"))
               ( (MEMBER !thefact diagnoses)
                 (PRINT !thefact " was a diagnosis to be investigated"))
               ( (MEMBER !thefact startingfacts)
                 (PRINT !thefact " was recorded from the outset"))
               (T (SETQ rules rulesfired)
                  (UNTIL (NULL rules)
                      (COND ( (if? !thefact (CAR rules))
                               (SETQ inif T)
                               (PRINT !thefact) (space)
                               (PRINT
                                  "was investigated to
                                  evaluate the certainty of:")
                               (writethensfired (CDDR (EVAL (CAR rules)))
                                                (CAR rules))))
                      (NEXTL rules))))
        (RETURN 'next?)))

(DE writethensfired (listthens therule) ;Called by why?
    (MAPCAR (LAMBDA (then) (space) (PRINT (NTH (1− then) listthens)))
            (GET therule 'thensfired)))

(DF use? !therules ;Called by the user himself
    (COND ( (NULL !therules)
             (space) (PRIN "rules fired: ") (writeseries rulesfired) (TERPRI)))
    (UNTIL (NULL !therules)
        (space)
        (COND ( (MEMBER (CAR !therules) rulesfired)
                 (PRINT (CAR !therules) " was fired"))
               ( (MEMBER (CAR !therules) rulebase)
                 (PRINT (CAR !therules) " was not fired"))
               (T (PRINT (CAR !therules) " does not exist"))))
        (NEXTL !therules))
    'next?)

(DE facts? () ;Called by the user himself
    (COND ((NULL factbase) (space) (PRINT "no recorded fact"))
          (T (writefacts factbase)))
    'next?)

(DE writefacts (list) ;Called by facts?
    (COND ((NULL list))
          (T (space)
             (PRINT (CAR list) " certain to degree: " (GET (CADR list) 'cf))
             (writefacts (CDDR list)))))
```

The comments that accompanied engine-program 2.2 relating to *interaction, treatment of multiple conclusions* and *detection of looping* remain valid for engine-program 2.3 (although some additional information on the third of these points will be supplied later on).

Representation of facts

Only three procedures can create new facts. Two of them cause facts to be recorded by calling createfact: these are the ask and deduce procedures which form part of the engine proper.

The third is the instantiation procedure (mycolog here) which directly affects factbase; in the case of mycolog this is limited to emptying factbase at the start. To initialise factbase to something else it would be necessary to adjust the instantiation program in accordance with the conventions described below.

The procedure updatefact, called by deduce, is the only one empowered to modify already existing facts (facts that are in the course of being evaluated). factbase is now organised as a list in which each odd numbered element is the text of a fact while each even numbered element is a label associated with the text of the preceding fact. For example, factbase could equal:

((rimose cap) g126 (thick gills) g125 (cortina) g124)

A new label is created for each new fact by a call to GENSYM (in createfact). At the same time, createfact sets the values of two properties, named cf and how, associated with each label. The first value, for the cf property, is that of the certainty factor currently assigned to the fact (it can change later on). The second value collects information that will allow type how? questions to be answered.

Before calling createfact, ask first carries out a check by means of procedure determined?: factbase only keeps track of a fact supplied by the user if the absolute value of the associated certainty factor is greater than the threshold (0.2).

Procedure deduce does not perform this check; the absolute value of the certainty factor associated with a fact can drop below the threshold, either at the time of its creation (when createfact is called by deduce) or when updates take place (updatefact is called exclusively by deduce). If a new rule is applied, the absolute value of the factor can rise above the threshold again; for example: the addition of a factor 0.15 and a factor 0.10 exceeds 0.2.

Absolute values which are *transiently* less than the threshold must be memorised because they need to be taken into account in the final evaluation of the fact; on the other hand, it is pointless (though deduce still does so) to remember absolute values which are less than the threshold *if they are final since weak certainty values are equated to the absence of information* when the premise is evaluated (in investigateit).

Conversely, it is natural to record negative certainty factors of absolute-value greater than the threshold (whether they come from the user or from deductions via negative attenuation coefficients) because it is significant to learn at the end of the diagnostic session that a certain fact has been assigned a degree of certainty equal to, say, -0.8 or -1.

The value of the how property of a fact label can take 2 forms according

to whether the fact is evaluated by a question or by deduction. In the first
case an expression of the form:

> (question <no. of the question which allowed the factor to be
> obtained>)

is stored. In the second case, an expression of the form:

> (<rule fired's label> <cumulative factor received> <last
> question no.>)

is stacked for each rule application.

For example, on the first call of how? in session 2.3, the how property of
the label associated with the rule (family boletaceae) has value: ((r9 42
8)(r18 30 3)).

Other conventions could have been employed for encoding the information relating to facts. The one that we have adopted lends itself easily to
later developments of the representation part of the generator's knowledge;
thus, properties other than cf and how could be attributed to fact labels.
Seen from this viewpoint, labels would identify *structured objects, each one
possibly* presenting various properties apart from cf and how: text of the
fact (or several texts for different parts of the fact), basic information to
facilitate responding later on to other types of questions (the names of the
rules in which the fact appears in the premise or conclusion, for instance),
information about how to turn the messages into relatively natural language,
information concerning relationships with other objects (*class to which it
belongs, meansof inheriting properties of other objects*).

Several procedures consult the fact base directly (factbase or depen-
dencies): flatfacts (which builds a list similar to factbase but replaces the
labels by the associated cf value), cfacquired?, howrule, how? and
writefacts.

Preparing replies to how? questions

The values of the how properties described in the preceding section are used
directly by writejustif (when how? is called). The response is either the
number of the question asked and the factor received in reply, or else the
names of the rules fired, each one accompanied by the current cumulative
factor and the number of the last question. When the user wants details he
must precede the wording of the fact by * (see the example in session 2.3). In
this case, further information is provided: the attenuation coefficient of each
rule fired and the factors associated with each of the facts involved in the
premise, which by then are bound to be final factors (see detailsfiring).

Procedure howrule consults descendantrules in order to get the
current rule's name. descendantrules is initialised to NIL by prologue and
then maintained as a stack by investigatethem. To satisfy the needs of
howrule alone the name of the current rule could have been passed as a
parameter from investigate down to howrule via deduce, createfact and
updatefact, but descendantrules is also used when jumping out because of
looping (see later on).

Recording facts which are definitively too uncertain

A technique was used in engine 2.2 to avoid restarting attempts at deduction concerning facts which have already been examined and finally been established as false. That technique has been extended here in a natural way: the tests on the binary value of the fact (established or not) have been replaced by tests on the absolute value of the certainty factor (exceeding the threshold or not), and a list of facts called notestablished is still maintained and consulted.

Thus a fact is placed in notestablished if the absolute value of the certainty factor is less than the threshold. In particular this list is used by investigateit and onenotestablished? so as to shorten the evaluation of certain premises. A variant could consist of finally storing all the facts examined (that is to say submitted to establishafact) in factbase, regardless of the outcome of the examination: NIL (because no rule applicable) or else any certainty factor. For this variant it suffices to alter establishafact, ask and notestablished? as follows (the variant gets rid of the peculiar list notestablished, simplifies the writing of establishafact and ask, does not really complicate notestablished?, and instead uses a label and the associated properties):

```
(DE establishafact (thefact) ;Called by retro and investigateit
    (PROG (rules cf)
        (COND ((SETQ cf (cfacquired? thefact)) (RETURN cf))
              ((MEMBER thefact ruleaims)
               (RETURN (doabandon thefact))))
        (SETQ rules (inthen thefact))
        (COND ((NULL rules) (RETURN (ask thefact))))
        (NEWL ruleaims thefact)
        (SETQ cf (investigatethem rules thefact))
        (NEXTL ruleaims)
        (RETURN cf)))

(DE notestablished (thefact) ;Called by establishafact and notestablished?
    (NOT (determined? (cfacquired? thefact))))

(DE ask (thefact) ;Called by establishafact
    (PROG (cfgiven)
        (SETQ nquestion (1+ nquestion))
        (TERPRI)
        (PRINT "QUESTION " nquestion ": certainty for " thefact
            " ? (0 if unknown)")
        (SETQ cfgiven (READ))
        (createfact thefact cfgiven NIL)
        (RETURN NIL)))
```

To accompany this variant, the test (NOT (cfacquired? thefact)) and the subsequent message in how? and why? can be suppressed.

In this variant, as in the original version, we have deliberately considered that the evaluation of the premise can be shortened once any one of the facts included in that premise was (i) already evaluated and (ii) assigned a factor in the range -0.2 to 0.2 (if the threshold is 0.2). Why not replace the second condition by: "assigned a factor *less than* 0.2"?

As already mentioned, in EMYCIN and its descendants the certainty degree of an elementary condition in a premise built on a fact which has a negative factor can itself be positive (because each elementary condition encapsulates an elementary fact in one or more functions which modulate its sense); when the rule language permits such a sophistication, it is normal to confine oneself to what we have implemented up to this point.

However, in the rule representation language allowed in engine-program 2.3 (and its predecessors) the premises are only conjunctions of elementary facts; hence, *within this more restricted framework,* the second kind of test can be adopted: the last versions of **establishafact** and ask are kept, and the following version of **notestablished?** is adopted:

> (DE **notestablished** (thefact) ;Called by establish a fact and one not established?
> (< (cfacquired? thefact) threshold))

Additional information concerning looping

The symbolic instantiation program below is designed to test the mechanism for detecting looping.

```
(DE trial ()
     (SETQ rulebase '(r1 r2 r3 r4 r5)
          r1 '(((f 3) (f 4)) 100 (f 1) (f 2)) r2 '(((f 6) (f 7)) 100 (f 4))
          r3 '(((f 8) (f 9)) 100 (f 4) (f 5)) r4 '(((f 10) (f 7)) 100 (f 4))
          r5 '(((f 11) (f 1)) 100 (f 9))
          factbase NIL
          diagnoses '((f 1)))
     (retro))
```

In the session reproduced below the system was run twice. The first execution runs without problem: as a result of the absolute certainty attached to (f 10), rule r4 (which happens to be the first one examined by **investigatethem**) is enough to arrive at the absolute certainty of (f 4). On the other hand, in the second session it is necessary to invoke r3 after r4 in order to evaluate (f 4) since f4 was unable to reach absolute certainty. To apply r3, (f 9) has to be evaluated and hence r5 invoked, which leads to evaluating (f 1). . . which is already being evaluated. The messages issued by do abandon permit easy analysis of the circumstances surrounding the looping. The reply to the question **how?** only provides information whose acquisition has been correctly completed.

(trial)

diagnosis (dia) or general information (gen) ?
dia

QUESTION 1: certainty for (f 3) ? (0 if unknown)

100

QUESTION 2: certainty for (f 10) ? (0 if unknown)
100

PROVISIONAL CONCLUSION, AFTER APPLYING RULE: r4
the cumulative certainty factor of (f 4) is 100

PROVISIONAL CONCLUSION, AFTER APPLYING RULE: r1
the cumulative certainty factor of (f 1) is 100

*** DIAGNOSIS: (f 1) is certain to degree 100

*** END OF THE DIAGNOSTIC SESSION
information ? (yes/no)
no

= *** bye

(trial)

diagnosis (dia) or general information (gen) ?
dia

QUESTION 1: certainty for (f 3) ? (0 if unknown)
100

QUESTION 2: certainty for (f 10) ? (0 if unknown)
50

PROVISIONAL CONCLUSION, AFTER APPLYING RULE: r4
the cumulative certainty factor of (f 4) is 50

QUESTION 3: certainty for (f 8) ? (0 if unknown)
100

QUESTION 4: certainty for (f 11) ? (0 if unknown)
100

*** ABANDONING ON ENCOUNTERING LOOPING ON: (f 1)
rule stack:
     r5
     r3
     r1
fact stack:

     (f 1)
     (f 9)
     (f 4)
     (f 1)

*** END OF THE DIAGNOSTIC SESSION
information ? (yes/no)
yes
= ask your questions
(facts?)
     (f 11) certain to degree: 100
     (f 8) certain to degree: 100

(f 4) certain to degree: 50
(f 10) certain to degree: 50
(f 3) certain to degree: 100
= next?
(how? * f 4)
    I used rule r4
    to conclude that: (f 4).
    This gave a cumulative certainty factor of: 50.
    since the rule's premise has 1 element
    certain to degree 50
    and the attenuation coefficient is 100.
    The question previously asked was 2.

= next?

The EMYCIN system watches for looping using the same technique as here.

When a loop is detected the system does not however abandon the session: instead it makes use of a recovery heuristic. In the example above, EMYCIN would prevent r5 being applied because it calls (f 1) which is currently under evaluation; having discovered that (f 1) can not be evaluated outside of the application of r5, the system would prevent r3 being applied too; then the system would invoke the last rule capable of reaching a conclusion about (f 4), namely r2, and would combine its possible contribution with that already obtained from rule r4. In addition, MYCIN can take advantage (by means of another heuristic equally worthy of discussion) of rules (called *self-referential rules*) of the form: IF F1 and F2 THEN F1 (see [BUCHANAN and SHORTLIFFE 1984], [FARRENY 1985]).

Miscellaneous remarks
Engines 2.1 and 2.2 stopped as soon as one of the possible diagnoses was established. Engine 2.3 only stops when all of the potential diagnoses have been evaluated. The difference between the systems is programmed by a slight modification to retro.

As soon as a fact receives a factor of -1 or $+1$ in the course of being evaluated the evaluation is stopped (line 7 of investigateit); in fact, the cumulative formulae are such that combining any factor other than -1 with 1 (or -1) gives 1 (or -1) again; thus the evaluation can be shortened. In the case where there is a conflict between one factor -1 and one factor $+1$, EMYCIN decides in favour of $+1$; engine 2.3 deviates from EMYCIN on this one very specific point: should a factor -1 be obtained no attempt is made to discover whether another factor could equal $+1$ and thereby prevail over it.

2.2.6   Asking questions in the course of the diagnostic session and correcting previous replies
Engine-program 2.3 makes a clear distinction between two phases of work, the diagnostic session being one, the information and explanation session

being the other; the user does not have the chance to ask questions during a diagnostic session. Furthermore, engine 2.3 does not allow a response to an earlier question to be rectified; this situation is all the more irritating since the user has no means of calling an immediate halt to the session while it is in progress. Some adjustments to engine 2.3 are now proposed which will allow (i) the diagnostic session to be interspersed with stages of information/explanation as the user pleases and (ii) erroneous replies to be revised.

Questions in the course of the diagnostic session
At what points is the user to be permitted to interrupt the progress of the search for a diagnosis and obtain explanations about its current state? In practice, a reasonable choice is to allow such interruptions only at those moments when the system addresses questions to the user. The use of these questions is illustrated in session 2.4 below.

If, in response to a request for a certainty factor, the user replies not with a number but with the code **I** (Information), a local information/explanation session is begun. The user can then resort to any of the 7 types of question: canbeused?, deducible?, rules?, facts?, why?, how?, and use?, *without abandoning the diagnostic session*: it is merely suspended and will continue normally if the user ends the local information session with the code **E** (End). The code **Q** (Quit) allows the diagnostic session to be stopped permanently (a new session being suggested).

Recall that the why? question type was defined as: "why *did* the system evaluate a certain fact?". Now, this primitive does not allow explanations to be obtained (even after the event) as to why *attempts* at evaluation were carried out by the engine. If the user responds to a question with the code ?, this code is interpreted as: "why is the system asking this question" or: "why is the system *looking for* the value of this fact?" (the one which is the object of the system's question to the user). The system indicates the rule it is trying to apply, the conclusion being pursued and the state of evaluation of the premise (in which the fact being asked about appears): in this way, it explains the stage of reasoning which directly motivated the question. If, after this reply, the user again uses the code ?, this is interpreted as: "and why is the system pursuing this conclusion?"; the system then explains the stage of reasoning which immediately preceded the one that motivated the question; and so on if desired.

Session 2.4 is obtained by jointly loading instantiation program 2.2 (very slightly modified as specified later on) and the extended engine-program 2.3 (the alterations will be described later on too: engine-program 2.4). It first illustrates the use of codes **I**, **F** and ? (possibly linked together: see the successive responses to questions 11). The meaning of code R will be indicated further on.

Session 2.4
(mycolog)

Diagnosis (D), Quit (Q) or Information/explanation (I) ?
D

*** START OF DIAGNOSTIC SESSION

QUESTION 1: certainty for (pliciform gills) ?
          (0 if unknown, H for Help)
80

PROVISIONAL CONCLUSION, AFTER APPLYING: r17
the cumulative certainty factor of (family cantharellaceae) is 40

*** DIAGNOSIS: (family cantharellaceae) is certain to degree 40

QUESTION 2: certainty for (dull spores) ?
          (0 if unknown, H for Help)
          90

          QUESTION 3: certainty for (thick gills) ?
                    (0 if unknown, H for Help)
60

PROVISIONAL CONCLUSION, AFTER APPLYING: r18
the cumulative certainty factor of (family boletaceae) is 30

QUESTION 4: certainty for (separating gills) ?
          (0 if unknown, H for Help)
25

QUESTION 5: certainty for (ochre spores) ?
          (0 if unknown, H for Help)
−50

QUESTION 6: certainty for (viscid cap) ?
          (0 if unknown, H for Help)
40

QUESTION 7: certainty for (no gills) ?
          (0 if unknown, H for Help)
70

QUESTION 8: certainty for (tubes) ?
          (0 if unknown, H for Help)
?   ← new code
I am interested in (tubes)
with a view to applying rule r9
so as to reach a conclusion about (family boletaceae)
which is a possible diagnosis

r9 comprises 3 condition(s) including (tubes)

I already know that (viscid cap) is certain to degree 40
I already know that (no gills) is certain to degree 70
I now need the certainty for: (tubes)

QUESTION 8: certainty for (tubes) ?
          (0 if unknown, H for Help)
50

PROVISIONAL CONCLUSION, AFTER APPLYING: r9

the cumulative certainty factor of (family boletaceae) is 42

*** DIAGNOSIS: (family boletaceae) is certain to degree 42

QUESTION 9: certainty for (bright coloured cap) ?
                    (0 if unknown, H for Help)
−100

QUESTION 10: certainty for (pink spores) ?
                    (0 if unknown, H for Help)
−100

QUESTION 11: certainty for (no stalk) ?
                    (0 if unknown, H for Help)
                    ?
                    I am interested in (no stalk)
                    with a view to applying rule r8
                    so as to reach a conclusion about (family tricholomataceae)
                    which is a possible diagnosis

                    r8 comprises 3 condition(s) including (no stalk)

                    I already know that (viscid cap) is certain to degree 40
                    I now need the certainty for: (no stalk)

                    QUESTION 11: certainty for (no stalk) ?
                                        (0 if unknown, H for Help)
I

*** START OF INFORMATION/EXPLANATION SESSION
Ask your questions (E for End)
(facts?)
        (pink spores) certain to degree: −100
        (bright coloured cap) certain to degree: −100
        (tubes) certain to degree: 50
        (no gills) certain to degree: 70
        (viscid cap) certain to degree: 40
        (ochre spores) certain to degree: −50
        (separating gills) certain to degree: 25
        (family boletaceae) certain to degree: 42
        (thick gills) certain to degree: 60
        (dull spores) certain to degree: 90
        (family cantharellacaea) certain to degree: 40
        (pliciform gills) certain to degree: 80
(how? ochre spores)

        you supplied the degree: −50
        in response to question: 5
E

*** END OF INFORMATION/EXPLANATION SESSION

QUESTION 11: certainty for (no stalk) ?
                    (0 if unknown, H for Help)
R

Which question no. do you want to start revising your replies from ?
55
The questions that can be revised are numbered from 1 to 10

QUESTION 11: certainty for (no stalk) ?
         (0 if unknown, H for Help)
         R
         Which question no. do you want to start revising your replies
         from ?
         5
         Hang on, let me recap . . .

         PROVISIONAL CONCLUSION, AFTER APPLYING: r17
         the cumulative certainty factor of (family cantharellaceae) is
         40

         *** DIAGNOSIS: (family cantharellaceae) is certain to degree
         40

         PROVISIONAL CONCLUSION, AFTER APPLYING: r18
         the cumulative certainty factor of (family boletaceae) is 30

         QUESTION 5: certainty for (ochre spores) ?
              (0 if unknown, H for Help)
0

QUESTION 6: certainty for (viscid cap) ?
         (0 if unknown, H for Help)
40

QUESTION 7: certainty for (no gills) ?
(0 if unknown, H for Help)
70

QUESTION 8: certainty for (tubes) ?
(0 if unknown, H for Help)
50

PROVISIONAL CONCLUSION, AFTER APPLYING: r9
the cumulative certainty factor of (family boletaceae) is 42

*** DIAGNOSIS: (family boletaceae) is certain to degree 42

QUESTION 9: certainty for (bright coloured cap) ?
         (0 if unknown, H for Help)
         −100

         QUESTION 10: certainty for (pink spores) ?
              (0 if unknown, H for Help)
−100

QUESTION 11: certainty for (tiny) ?
         (0 if unknown, H for Help)
I

*** START OF INFORMATION/EXPLANATION SESSION

Ask your questions (E for End)
(facts?)
     (pink spores) certain to degree: −100
     (bright coloured cap) certain to degree: −100
     (tubes) certain to degree: 50
     (no gills) certain to degree: 70
     (viscid cap) certain to degree: 40
     (separating gills) certain to degree: 25
     (family boletaceae) certain to degree: 42
     (thick gills) certain to degree: 60
     (dull spores) certain to degree: 90
     (family cantharellaceae) certain to degree: 40
     (pliciform gills) certain to degree: 80
E

*** END OF INFORMATION/EXPLANATION SESSION
QUESTION 11: certainty for (tiny) ?
                      (0 if unknown, H for Help)
?

I am interested in (tiny)
with a view to applying rule r10
so as to reach a conclusion about (genus marasmius)

r10 comprises 4 condition(s) including (tiny)

I now need the certainty for: (tiny)

QUESTION 11: certainty for (tiny) ?
             (0 if unknown, H for Help)
?

I am interested in (genus marasmius)
with a view to applying rule r27
so as to reach a conclusion about (family marasmiaceae)
which is a possible diagnosis

r27 comprises 1 condition(s) including (genus marasmius)

I now need the certainty for: (genus marasmius)

QUESTION 11: certainty for (tiny) ?
             (0 if unknown, H for Help)

Nothing more to say
QUESTION 11: certainty for (tiny) ?
             (0 if unknown, H for Help)
−80

QUESTION 12: certainty for (rimose cap) ?
(0 if unknown, H for Help)
−50

QUESTION 13: certainty for (black spores) ?
(0 if unknown, H for Help)
30

QUESTION 14: certainty for (cap not exposed) ?
           (0 if unknown, H for Help)
?

I am interested in (cap not exposed)
with a view to applying rule r20
so as to reach a conclusion about (family naucoriaceae)
which is a possible diagnosis

r20 comprises 2 condition(s) including (cap not exposed)

I already know that (black spores) is certain to degree 30
I now need the certainty for: (cap not exposed)

QUESTION 14: certainty for (cap not exposed) ?
           (0 if unknown, H for Help)
I

*** START OF INFORMATION/EXPLANATION SESSION
Ask your questions (E for End)
(rules? r20)

rule: r20
     IF:
     (black spores)
     (cap not exposed)

     THEN (40):
     (genus panaeolus)
     (family naucoriaceae)
E

*** END OF INFORMATION/EXPLANATION SESSION
QUESTION 14: certainty for (cap not exposed) ?
           (0 if unknown, H for Help)
R

Which question no. do you want to start revising your replies from ?
13
Hang on, let me recap . . .

PROVISIONAL CONCLUSION, AFTER APPLYING: r17
the cumulative certainty factor of (family cantharellaceae) is 40

*** DIAGNOSIS: (family cantharellaceae) is certain to degree 40

PROVISIONAL CONCLUSION, AFTER APPLYING: r18
the cumulative certainty factor of (family boletaceae) is 30

PROVISIONAL CONCLUSION, AFTER APPLYING: r9
the cumulative certainty factor of (family boletaceae) is 42

*** DIAGNOSIS: (family boletaceae) is certain to degree 42

QUESTION 13: certainty for (black spores) ?
           (0 if unknown, H for Help)
−90

QUESTION 14: certainty for (cupuloform cap) ?
            (0 if unknown, H for Help)
30

QUESTION 15: certainty for (gelatinous coat on cap) ?
            (0 if unknown, H for Help)
−80

QUESTION 16: certainty for (granular flesh) ?
            (0 if unknown, H for Help)
−100

*** END OF DIAGNOSTIC SESSION

Diagnosis (D), Quit (Q) or Information/explanation (I) ?
I

*** START OF INFORMATION/EXPLANATION SESSION
Ask your questions (E for End)
(facts?)
     (granular flesh) certain to degree: −100
     (gelatinous coat on cap) certain to degree: −80
     (cupuloform cap) certain to degree: 30
     (black spores) certain to degree: −90
     (rimose cap) certain to degree: −50
     (tiny) certain to degree: −80
     (pink spores) certain to degree: −100
     (bright coloured cap) certain to degree: −100
     (tubes) certain to degree: 50
     (no gills) certain to degree: 70
     (viscid cap) certain to degree: 40
     (separating gills) certain to degree: 25
     (family boletaceae) certain to degree: 42
     (thick gills) certain to degree: 60
     (dull spores) certain to degree: 90
     (family cantharellaceae) certain to degree: 40
     (pliciform gills) certain to degree: 80
(use?)
     rules fired: r9 r18 r17
E

***END OF INFORMATION/EXPLANATION SESSION

Diagnosis (D), Quit (Q) or Information/explanation (I) ?
Q
= bye

Correcting replies in the course of a diagnostic session
Right from the earliest versions (1974) the EMYCIN system allowed the
user to cause a part of the diagnostic session to be revised. The user would
give the number of a question to which he had replied incorrectly. The

system would reconstitute the state of the fact base which existed just prior to when the incriminating response was accepted and would recommence the session starting from that stage. The modifications to engine-program 2.3 described hereafter, which allowed sessions 2.3 and 2.4 to take place as they did, offer the user two variants for correcting replies.

In the first variant the user answers a request for a certainty factor with the code **R** (Revision). The system then asks for the number of the question whose reply is to be revised. It retains the responses obtained before that question but forgets all the replies thereafter. This method is adapted to the cases where correcting an answer is presumed to require the correction of all the replies which follow, or the majority of them. At present, altering a reply makes certain subsequent questions inappropriate. Two revisions of this kind are carried out in the course of session 2.4.

We should note in passing that session 2.4 also illustrates how the evaluation of a rule premise can be shortened if *any* of the facts comprising the premise is already permanently listed as *too uncertain to permit a rule application*. Recall that in engine-program 2.3, as in EMYCIN, *too uncertain to permit a rule application* means: having a factor between -0.2 and $+0.2$ (if threshold$=0.2$). We have also indicated how and why the program could be adapted to interpret *too uncertain to permit a rule application* as: less than $+0.2$. The first part of the session (questions 1 to 11) progresses without any response being challenged; after question 5, a factor of -50 (equivalent to -0.5) has been assigned to (ochre spores) by the user; at question 11 the system requests the factor to be associated with (no stalk) with the intention of applying rule r8.

As the session continues the user corrects the certainty factor assigned to (ochre spores): he substitutes 0 for -50; it can be seen that there is no longer a question about (no stalk) in the rest of the session. In fact, (ochre spores) appears as the third condition in the premise of r8; but when r8 is examined the fact (ochre spores) is assigned to 0 which allows the rule to be dismissed without asking a question about (no stalk).

In the second variant, the user responds with the code **SR** (Selective Revision). The system then asks for the list of question numbers whose answers must be reconsidered. However, the other (preceding or subsequent) responses remain valid. In this case the system restarts the diagnostic session from the beginning without repeating any questions not designated for revision: the replies to such questions are all kept as they are. Of course, it is possible that if a given reply had been correct from the outset then a given question, whose answer is still kept here, would not have been asked. It is even possible that correcting a particular response is of no further interest once some other response has been corrected beforehand. In session 2.5 it should be noticed that the SR reply to question 11, in place of the R reply to the same question in session 2.4, reduces the subsequent number of questions (questions 6, 7, 8 and 10 are not put a second time). In order to shorten the illustration we have chosen to use the **Q** code in response to question 12.

Session 2.5
(mycolog)

Diagnosis (D), Quit (Q) or Information/explanation (I) ?
D

*** START OF DIAGNOSTIC SESSION
QUESTION 1: certainty for (pliciform gills) ?
          (0 if unknown, H for Help)
80

PROVISIONAL CONCLUSION, AFTER APPLYING: r17
the cumulative certainty factor of (family cantharellaceae) is 40

*** DIAGNOSIS: (family cantharellaceae) is certain to degree 40

*** QUESTION 2: certainty for (dull spores) ?
          (0 if unknown, H for Help)
90

QUESTION 3: certainty for (thick gills) ?
          (0 if unknown, H for Help)
60

PROVISIONAL CONCLUSION, AFTER APPLYING: r18
the cumulative certainty factor of (family boletaceae) is 30

QUESTION 4: certainty for (separating gills) ?
          (0 if unknown, H for Help)
25

QUESTION 5: certainty for (ochre spores) ?
          (0 if unknown, H for Help)
−50

QUESTION 6: certainty for (viscid cap) ?
          (0 if unknown, H for Help)
40

QUESTION 7: certainty for (no gills) ?
          (0 if unknown, H for Help)
70

QUESTION 8: certainty for (tubes) ?
          (0 if unknown, H for Help)
50

PROVISIONAL CONCLUSION, AFTER APPLYING: r9
the cumulative certainty factor of (family boletaceae) is 42

*** DIAGNOSIS: (family boletaceae) is certain to degree 42

QUESTION 9: certainty for (bright coloured cap) ?
          (0 if unknown, H for Help)
—30

QUESTION 10: certainty for (pink spores) ?
          (0 if unknown, H for Help)

−100

QUESTION 11: certainty for (no stalk) ?
            (0 if unknown, H for Help)
SR

Give the list of question numbers you want to go back to
(5 9)

Hang on, let me recap . . .

PROVISIONAL CONCLUSION, AFTER APPLYING: r17
the cumulative certainty factor of (family cantharellaceae) is 40

*** DIAGNOSIS: (family cantharellaceae) is certain to degree 40

PROVISIONAL CONCLUSION, AFTER APPLYING: r18
the cumulative certainty factor of (family boletaceae) is 30

QUESTION 5: certainty for (ochre spores) ?
            (0 if unknown, H for Help)

PROVISIONAL CONCLUSION, AFTER APPLYING: r9
the cumulative certainty factor of (family boletaceae) is 42

*** DIAGNOSIS: (family boletaceae) is certain to degree 42

QUESTION 9: certainty for (bright coloured cap) ?
            (0 if unknown, H for Help)
−100

QUESTION 11: certainty for (tiny) ?
            (0 if unknown, H for Help)
−80

QUESTION 12: certainty for (rimose cap) ?
            (0 if unknown, H for Help)
Q

*** HALTED ON YOUR INSTRUCTION, AFTER QUESTION 12
*** END OF DIAGNOSTIC SESSION

Diagnosis (D), Quit (Q) or Information/explanation (I) ?
Q
= bye

Engine-program 2.4

To allow the user to do the above we start from engine 2.3. Modifications in already existing procedures are underlined in the body. The names of new procedures are also underlined. Procedures prologue, epilogue and flat-facts disappear. Only one change is made to the instantiation program: the final call of (retro) is replaced by (antiretro).

```
(DE antiretro () ;Called by the instantiation program
;cfagain, startingfacts and factbase are global variables
    (PROG (reply)
        (TERPRI)
        (PRINT "Diagnosis (D), Quit (Q) or Information/explanation (I) ?")
        (SETQ reply (READ))
        (COND ( (EQ reply 'Q) (RETURN 'bye))
               ( (EQ reply 'I) (TERPRI)
                (PRINT "*** START OF INFORMATION/EXPLANATION
                        SESSION")
                (PRINT "Ask your questions (E for End)")
                (info))
               ( (EQ reply 'D) (TERPRI)
                (PRINT "*** START OF DIAGNOSTIC SESSION")
                (SETQ cfagain NIL startingfacts factbase)
                (TAG revision (retro))))
        (antiretro)))

(DE info () ;Called by antiretro and ask
    (PROG (request)
        (SETQ request (READ))
        (COND ( (EQ request 'F) (TERPRI)
                (PRINT "*** END OF INFORMATION/EXPLANATION
SESSION"))
               (T (EVAL request) (info)))))

(DE retro ()
;diagnoses, startingfacts, factbase, rulesfired, notestablished, ruleaims,
;nquestion, descendantrules, threshold and cfsreplied are global variables
;possibles is declared in retro
    (PROG (possibles afact cf result)
    (SETQ possibles diagnoses rulesfired NIL notestablished NIL ruleaims
          NIL nquestion 0 ruledescendants NIL threshold 20 cfsreplied NIL)
        (TAG abandon
            (UNTIL (NULL possibles)
                (SETQ afact (NEXTL possibles) cf (establishafact afact))
                (COND ((determined? cf) (writediagno afact) (SETQ result
                T))))
            (COND ((NULL result) (writenodiagno))))
        (PRINT "*** END OF DIAGNOSTIC SESSION")))

(DE ask (thefact) ;Called by establishafact
    (PROG (reply stage)
        (SETQ nquestion (1+ nquestion)) (COND ((AND cfagain (CAR cfagain))
        (NEXTL cfagain reply))
               (T (NEXTL cfagain)
                (TERPRI)
                (SETQ stage -1)
                (UNTIL
                    (AND (NUMBERP reply) (<= reply 100) (>= reply -100))
                    (PRINT "QUESTION " nquestion ": certainty for " thefact " ?")
```

```
                              (space) (PRINT "(0 if unknown, H for Help)")
                              (SETQ reply (READ))
                              (COND ((EQ reply '?) (seestage thefact (SETQ stage
                              (1+ stage))))
                                    ((EQ reply 'I)
                                    (PRINT
                                       "*** START OF INFORMATION/
                                       EXPLANATION SESSION")
                                    (PRINT "Ask your questions (E for End)")
                                    (info))
                                    ((EQ reply 'R) (revision))
                                    ((EQ reply 'SR) (selectrevis))
                                    ((EQ reply 'Q) (RETURN (quit)))
                                    ((EQ reply 'H) (help))))))
        (NEWL cfsreplied reply)
        (COND ((determined? reply) (createfact thefact reply NIL) (RETURN
        reply)))
        (NEWL notestablished thefact)
        (RETURN NIL)))

(DE seestage (fact n) ;Called by ask
    (PROG (theconditions rule limit)
        (TERPRI)
        (SETQ limit (LENGTH ruleaims))
        (COND ((>= n limit) (RETURN (PRINT "Nothing more to say")))
              ((> n 0) (SETQ fact (CAR (NTHCDR (1-n) ruleaims)))))
        (SETQ rule (CAR (NTHCDR n descendantrules)))
        (PRINT "I am interested in " fact)
        (PRINT "with a view to applying rule " rule)
        (PRINT "so as to reach a conclusion about " (CAR (NTHCDR n
        ruleaims)))
        (COND ((= n (1- limit)) (PRINT "which is a possible diagnosis")))
        (SETQ theconditions (CAR (EVAL rule)))
        (TERPRI)
        (PRINT rule " comprises " (LENGTH theconditions)
              " condition(s) including " fact)
        (TERPRI)
        (UNTIL (EQUAL fact (CAR theconditions))
              (PRINT "I already know that " (CAR theconditions)
                     " is certain to degree " (cfacquired? (CAR theconditions)))
              (NEXTL theconditions))
        (PRINT "I now need the certainty for: " fact)
        (TERPRI)))

(DE revision () ;Called by ask
    (PROG (backto)
        (TERPRI)
        (PRINT
            "Which question no. do you want to start revising your replies
            from ?")
        (SETQ backto (READ))
```

```lisp
(COND ( (OR (NOT (NUMBERP backto)) (< backto 1) (>= backto
nquestion))
        (RETURN (wordnum))))
(PRINT "Hang on, let me recap . . .")
(SETQ factbase startingfacts
      cfagain (REVERSE (NTHCDR (" nquestion backto)
      cfsreplied)))
(EXIT revision (retro))))
```

```lisp
(DE wordnum () ;Called by revision and sift
   (PRINT "The questions that can be revised are numbered from 1 to "
          (1- nquestion)))
```

```lisp
(DE selectrevis () ;Called by ask
;selectrevis and revision could be combined into a single procedure
;(with a parameter)
   (PROG (tobecorrected)
      (TERPRI)
      (PRINT "Give the list of question numbers you want to go back to")
      (SETQ tobecorrected (READ))
      (COND ((NOT (LISTP tobecorrected)) (RETURN (PRINT "A list
please"))))
      (PRINT "Hang on, let me recap . . .")
      (remake startingfacts)
      (SETQ factbase startingfacts
            cfagain (sift tobecorrected (REVERSE cfsreplied)))
      (EXIT revision (retro))))
```

```lisp
(DE quit () ;Called by ask
   (TERPRI) (PRINT "***HALTED ON YOUR INSTRUCTION, AFTER
QUESTION " nquestion)
   (EXIT abandon))
```

```lisp
(DE sift () ;Called by selectrevis
   (COND ((NULL list1) list2)
         ((OR (< (CAR list1) 1) (>= (CAR list1) nquestion)) (RETURN
(wordnum)))
         (T (sift (CDR list1)
                  (SUBST NIL (NTH (1- (CAR list1)) list2) list2)))))
```

```lisp
(DE help () ;Called by ask
   (TERPRI) (PRINT "Replies expected:") (space)
   (PRINT "An integer between -100 and +100) (space)
   (PRINT "? for local explanation, I for interrogation session") (space)
   (PRINT "R for general revision, SR for selective revision")
   (PRINT "Q for quit everything, H for this help"))
```

```lisp
(DE startingcf? (thefact) ;Called by how?
   (PROG (temp)
      (COND ((SETQ temp (MEMBER thefact startingfacts)) (GET (CADR
temp) 'cf)))))
```

To allow the user's replies to be revised, the initial state of the fact base
has to be restored. To accomplish this, the identifier **startingfacts** will be set

in a simpler fashion than in the previous version: it will be given a copy of the initial fact base. There is no point in recording the certainty factors of the initial facts in **startingfacts** since *these factors cannot change*; there is no reason to restore them, for they are always available via the **cf** property of **facts**. The **startingcf?** procedure is modified as a consequence.

The certainty factors provided by the user in response to the system's questions are stacked in **cfsreplied** (see the end of **ask**). If code R is selected, **ask** calls **revision**; **revision** requests the number of the first question to which the user wants to return and, having checked that the answer is valid, re-initialises **factbase** (through **startingfacts**). **revision** then sets **cfagain** to the list of the first to the last of the user's valid replies; finally **revision** takes an escape route (also called **revision**, declared in **antiretro**) by calling **retro**. This causes a diagnostic session to be set off again; but **ask** will not put any questions as long as **cfagain** is not empty (see line 4 of **ask**, the first test in the AND); **cfagain** is decremented on each call to **ask**. If code SR is selected, **selectrevis** requests the list of responses the user wants to go back over and, having checked that the answer is valid, re-initialises **factbase**; **selectrevis** then sets **cfagain** to the list of all the user's responses, but replacing the replies to be corrected with **NIL**; finally **selectrevis** performs the same escape as **revision**. The second test in line 4 of **ask**, dealing with (CAR cfagain), ensures that only those questions indicated by the user are asked again.

The rest of the procedures can easily be analysed in the light of sessions 2.4 and 2.5.

2.3   PLANNING-TYPE EXPERT SYSTEM GENERATORS

Certain generators, including NOAH [SACERDOTI 1977], TROPIC [LATOMBE 1977] and ARGOS-II [FARRENY 1980] [PICARDAT 1985], have been specifically conceived to generate *action plans* intended for execution by human agents or by robots. Here we shall consider *linear* plans only, which means consisting of only one sequence of stages as in TROPIC and ARGOS-II, and not plans with *parallel branches* (which would permit switching to alternative sequences of stages, as in NOAH).

2.3.1   Example: a database for generating plans
Fig. 2.1 proposes a symbolic rule base, intended to represent expertise in planning for a particular, but unspecified, application domain.

> **rule-base**
R1:   P1 $<--$ A1, A2
R2:   P1 $<--$ A3
R3:   P2 $<--$ P1, A4, P3
R4:   P3, F1 $<--$ A5, P4
R5:   P3, F2 $<--$ P5
R6:   P3 $<--$ A4, P6
R7:   P4, F2, F3 $<--$ A2

> **fact-base** (facts established): set {F3, F4}

> **problem-base** (facts to be established): list (P2)

> **terminal-actions-table** (primitive problems)

symbol	facts added	facts removed
A1	F1, F2	none
A2	none	F1
A3	F1	none
A4	none	none
A5	F2	F4

Fig. 2.1

The rule named R1, for instance, will be interpreted as: "in order to solve the *problem* P1, simply execute the *terminal action* A1 and then the *terminal action* A2"; rule R4 will be interpreted as: "in order to solve the *problem* P3, given that the *fact* F1 is established, simply execute the *terminal action* A5 and then solve the *problem* P4". A concrete example of R4 might be: "to get to Perpignan (P3), given that we are at Orly-Ville (F1), just go from Orly-Ville to Paris-Austerlitz by Orly-Rail (A5) and then go from Paris-Austerlitz to Perpignan by train (P4).

Note: the interpretation that we have just proposed for rule R4 suggests a mode of invocation which derives partly from backward-chaining and partly from forward-chaining; backward-chaining because the rule is invoked when fact P3 is to be established (i.e. problem P3 is to be solved), forward-chaining because the rule is invoked when fact F1 is established. In fact, let us agree that rules are invoked by examining their left-hand sides; each element in the left-hand side is a pattern (reduced to a simple symbol in the example, but it could be a more complex structure). It is decided that the first pattern will refer to a special part of the fact base containing only facts to be established (which is to say problems to be solved); the other patterns, if any, in each rule will refer to the part of the fact base which, by convention, contains only facts that have been established. When the invocation of a rule is *simultaneously* conditioned by comparison with the facts to be established and the facts that have been established, we talk about *mixed-chaining*.

For each of the *problems* P1, P2, P3 and P4, one or more rules exist which are capable of solving the problem in an **ordered conjunction** of *action executions* and/or of solutions to other problems. The *terminal actions* A1, A2, A5, etc. . . . are *primitive problems*: this means that there is no need to solve them by recourse to rules like those which break down P1, P2, P3 and P4; the *terminal* actions stand for elementary actions, immediately execu- table by the human agent or robot considered in the application (basic movements of tasks), while *problems* require prior resolution (or analysis). An *action plan* is a series of terminal actions (no problems); for example: A3 A4 A5 A2.

When a terminal action (from here on called simply: action) is inserted into a plan being elaborated, the *effects* of this action must be represented by adding and retracting facts; for instance, having inserted the action "go from Orly–Ville to Paris–Austerlitz by Orly–Rail" into the plan, the fact which says we are at Orly–Ville has to be suppressed and the fact which says we are at Paris–Austerlitz has to be added. Adding a fact is interpreted as: establish the fact, while retracting a fact is interpreted as: the fact is not or is no longer established. A single action can appear in several rules; here, so as not to complicate writing the rules, the effects of each action are collected in a *terminal action table*, though rule R4 could have been directly coded as:

P3, F1 $<--$ INSERT–STAGE A5, ADD F2, RETRACT F4, P4

To make the example easier to follow, we have used the notation: Pi for the problems (or facts to be established), Aj for the actions and Fk for the (established) facts. However, the distinctions implied by the prefixes P, A and F are not vital; on the one hand, the first element of a rule's left-hand side is by convention a problem and the other elements, if they are present, are facts; on the other hand, an element of the right-hand side of a rule is an action if there is a line in the terminal actions table corresponding to it, otherwise it is a problem.

2.3.2   Outline of a nonmonotonic, mixed-chaining, depth first and tentative regime engine

Algorithm 2.3 below gives the outline for an inference engine using a knowledge base along the lines of the one proposed in Fig. 2.1. The engine is started by calling the main procedure, planretro. The keyword VARIABLE precedes the declarations of local variables in addition to the procedures' parameters; the keyword RETURN causes a procedure to be exited, returning the value of the expression which follows. The variables fact-base, problem-base, rule-base and terminal-actions-table are global to all procedures. They are presumed to be already initialised when the engine is set off.

To make it easier to understand how the algorithm works, it is a good idea to follow the progress of the commented example supplied further on.

Algorithm 2.3
```
1    PROCEDURE planretro
2        VARIABLE plan-base
3        plan-base <--- {}           (empty list)
4        IF solve(problem-base, fact-base, plan-base, rule-base) = "failure"
             THEN
5            WRITE ["no plan for the series of problems submitted"]
6            RETURN "failure"
7        ENDIF
8        WRITE ["plan for the series of problems submitted"]
9        display plan
10       RETURN "success"
```

```
1     PROCEDURE solve(problems, facts, plan, series)
2         VARIABLE cycle-result
3         IF problem-base = {} THEN RETURN "success"
4         cycle-result <--- execute-a-cycle(head of problem-base, series)
5         IF cycle-result = "failure" THEN RETURN "failure"
6         IF solve(problem-base, fact-base, plan-base, rule-base) = "success"
              THEN RETURN "success"
7         IF cycle-result = "primitive" THEN RETURN "failure"
8         problem-base <--- problems
9         fact-base <--- facts
10        plan-base <--- plan
11        RETURN solve(problems, facts, plan, tail of cycle-result)
```

```
1     PROCEDURE execute-a-cycle(the-problem, the-rules)
2         VARIABLE a-rule
3         IF the-problem is present in terminal-actions-table THEN
4             suppress the head of problem-base (which is precisely
5             the-problem) place the-problem at the tail of plan-base
6             carry out the additions to and retractions from fact-base
                  corresponding to the-problem according to terminal-actions-table
7             RETURN "primitive"
8         ENDIF
9         IF the-rules = {} THEN RETURN "failure"
10        a-rule <--- the head of the-rules
11        IF the left-hand side of a-rule is
                      compatible with the-problem and fact-base THEN
12            suppress the head of problem-base
13            place the elements of the right-hand side of a-rule into the head of
                  problem-base (maintaining their order)
14            RETURN the-rules 15
15        ENDIF
16        RETURN execute-a-cycle(the-problem, tail of the-rules)
```

General comments about algorithm 2.3

At the outset, all of the problems in problem-base are to be solved;
problem-base is a list; one plan is sought which *successively* solves all of the
problems in problem-base; one by one they are passed to execute-a-cycle
in the order given by the list; execute-a-cycle looks for a rule whose first
pattern is *compatible* (in this case, compatibility between a pattern and an
element of problem-base or fact-base is simply equality) with the current
problem, which is the *first* one in the list. When the left-hand side of a rule is
compatible with fact-base and rule-base (the overall compatibility of the
left-hand side is defined as the conjunction of compatibilities between
patterns and elements of the bases concerned), the current problem is
replaced by the ordered list of (possibly primitive) problems forming the
right-hand side. So problem-base ends up being handled as a stack and the
engine obeys a *depth first* type strategy.

If problem-base becomes empty, the engine stops: the initial series of
problems has been solved and the current state of plan-base is displayed. If
problem-base is not empty but no rule is compatible with the situation
created (characterised by the current state of fact-base and problem-

base), the engine systematically tries backtracking, meaning that it tries to apply a new rule to the last non-primitive problem retracted from **problem-base**, and so on. Hence the engine uses a *tentative regime* type system.

The knowledge expression language that algorithm 2.3 has to interpret is different in an important respect from the one interpreted by the generators studied in earlier sections: the "facts removed" column in the terminal actions table allows conclusions which have previously been established to be retracted (this will take place when the terminal action to which it is associated is inserted into the plan being elaborated). Since it can *take back* already *established facts*, the engine represented by algorithm 2.3 is said to be *nonmonotonic*. Note that even if there were no "facts removed" column but only a "facts added" column this engine would still be nonmonotonic. In fact, when algorithm 2.3 carries out a backtrack (towards a situation characterised by the states of **fact-base** and **problem-base**) it also restores the state of plan-base and completely removes the effects of any terminal actions that may have been retracted from the plan; thus it is able to take back additions of facts, which means it can take back facts that have been established.

Progress of algorithm 2.3 on an example
Fig. 2.2 shows an AND/OR graph representing the rules chained when

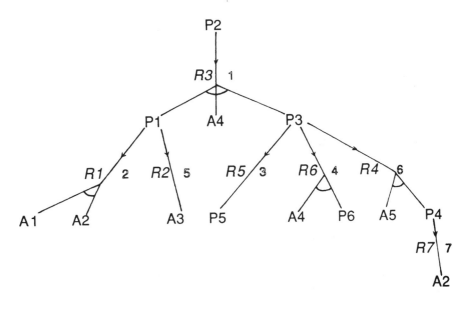

Fig. 2.2

algorithm 2.3 is applied to the knowledge base of Fig. 2.1. It shows, for example, that the first rule to be applied (marked 1) is *R3*; it decomposes P2 into P1, A4 and P3.

Fig. 2.3 shows in tabular form how **fact-base**, **problem-base** and **plan-base** evolve as the algorithm progresses. The first line is the initial state. The lines numbered from 1 to 7 show the results of the 7 rule applications performed; when a rule is applied, **problem-base** alone is affected. The lines marked "actions popped" show the results of finding terminal actions at the head of **problem-base**; an action (or several, one after the other) is retracted from **problem-base** and put into the head of **plan-base** while the corresponding additions and retractions are carried out in **fact-base**. Lines marked "backtrack" show where situations that have become blocked are exited: the algorithm has found no rule (or no more rules) that can be applied to the head of **problem-base** as it was in the preceding line; it restores the situation (**fact-base**, **problem-base** and **plan-base** states) which existed before the last rule application.

The algorithm ceases when **problem-base** becomes empty; the plan finally generated is: (A3 A4 A5 A2).

	fact-base	problem-base	plan-base
start	F3 F4	P2	{}
1 (after R3)	F3 F4	P1 A4 P3	{}
2 (after R1)	F3 F4	A1 A2 A4 P3	{}
actions popped	F2 F3 F4	P3	A1 A2 A4
3 (after R5)	F2 F3 F4	P5	A1 A2 A4
backtrack	F1 F2 F3 F4	P3	A1 A2 A4
4 (after R6)	F2 F3 F4	A4 P6	A1 A2 A4
actions popped	F2 F3 F4	P6	A1 A2 A4 A4
backtrack	F2 F3 F4	P3	A1 A2 A4
backtrack	F3 F4	P1 A4 P3	{}
5 (after R2)	F3 F4	A3 A4 P3	{}
actions popped	F1 F3 F4	P3	A3 A4
6 (after R4)	F1 F3 F4	A5 P4	A3 A4
actions popped	F1 F2 F3	A2	A3 A4 A5
7 (after R7)	F1 F2 F3	A2	A3 A4 A5
actions popped	F1 F2 F3 F4	{}	A3 A4 A5 A2

Fig. 2.3

Comments on the code of algorithm 2.3

The procedure **solve** tries to reduce the problem at the head of **problem-base** (call this problem PB) by calling **execute-a-cycle**. The parameter -the-rules in **execute-a-cycle** is passed the list of rules stored in the parameter of **solve** called **series**. When solve is called by **planretro**, **series** receives the whole rule base; this is also the case at certain times when **solve** calls itself (instruction 6); at other times when **solve** calls itself (instruction 11), **series** is only passed a sub-list of **rule-base**.

There are four cases to be considered:

— 1st case: **execute-a-cycle** returns "failure" because PB is not a primitive problem and cannot be broken down by any of the rules available in the list series. The current procedure solve then exits and returns "failure" to the next level up.

   If execute-a-cycle does not return "failure" then PB is either a primitive problem (terminal action), or else a problem which can be decomposed by one of the rules in the list series. In both of these circumstances solve is immediately called with arguments problem-base, fact-base and plan-base as they are after execute-a-cycle has dealt with PB, while the whole of the rule base is passed as the fourth parameter. From now on this call to solve is denoted A.
— hence 2nd case: if call A returns "success", the current invocation of solve returns "success" to the next level up.

   Otherwise, processing differs according to whether execute-a-cycle recognised PB as a primitive problem or attempted to break it down.
— 3rd case: if execute-a-cycle has returned "primitive" (because PB is a terminal action) then the failure of A's call to solve means that the arrival of PB at the head of problem-base must be undone; the current invocation of solve therefore returns "failure" to the next level up.
— 4th case: if PB is not a terminal action then PB can be decomposed. Hence execute-a-cycle has set off a rule to break down PB, placed the result of this at the head of problem-base and returned the sub-list of rule-base that begins with the rule triggered. In this case, the failure of A's call to solve may stem from the introduction of PB's subproblems. An attempt must therefore be made to find another way of decomposing PB: the context (states of problem-base, fact-base and plan-base) which existed before execute-a-cycle was called is now restored and a new call of solve is set off. The last argument of this call is the list of rules returned by execute-a-cycle, minus the rule already tried.

2.3.3   Basic LISP program for a nonmonotonic, mixed-chaining, depth first and tentative regime planning engine

Program 2.5 below is the result of directly implementing algorithm 2.3. There is only one notable difference: the engine can interpret rules whose left-hand side's first element is simply * (asterisk); in this case the engine does not compare any element in the left-hand side with the head of problembase: instead it goes directly to the comparison of the other elements in the left-hand side with factbase. These rules lend themselves to a purely *forward-chaining* type invocation. Thus the mixed-chaining engine 2.5 allows forward-chaining as well as backward-chaining as special cases.

Engine-program 2.5
```
(DE planretro () ;main procedure of the engine
    (PROG (planbase)
        (TERPRI)
        (COND ( (NULL (resolve problembase factbase planbase rulebase))
                (TERPRI) (PRINT "NO PLAN FOR THE SERIES OF PROBLEMS
                        SUBMITTED")
```

```
                    (RETURN 'failure)))
            (TERPRI)
            (PRINT "PLAN FOR THE SERIES OF PROBLEMS SUBMITTED:")
            (writeplan (REVERSE planbase))
            (RETURN 'success)))

(DE writeplan (steps) ;Called by planretro
    (MAPCAR (LAMBDA (×) (PRINCH '| | 12) (PRINT ×)) steps))

(DE solve (problems facts plan series) ;Called by planretro and solve
    (PROG (cycleresult)
            (COND ((NULL problembase) (RETURN T)))
            (SETQ cycleresult (executeacycle (CAR problembase) series))
            (COND ((NULL cycleresult) (RETURN NIL))
                    ((solve problembase factbase planbase rulebase)
                     (RETURN T))
                    ((EQ cycleresult 'primitive) (RETURN NIL)))
            (SETQ problembase problems factbase facts planbase plan)
            (writebacktrack)
            (solve problems facts plan (CDR cycleresult))))

(DE writebacktrack () ;Called by solve
    (PRINT "BACKTRACK") (PRINT "facts: " factbase)
    (PRINT "problems: " problembase) (PRINT "plan: " (REVERSE plan))
    (TERPRI))

(De executeacycle (theproblem therules) ;Called by solve and executeacycle
    (PROG (arule)
            (COND ( (ASSQ theproblem terminalactionstable)
                        (NEXTL problembase) (NEWL planbase theproblem)
                        (effects theproblem)
                        (writeinsertion theproblem)
                        (RETURN 'primitive))
                    ( (NULL therules) (TERPRI)
                        (PRIN "no (more) rules applicable to " theproblem" ===> ")
                        (RETURN NIL)))
            (SETQ arule (CAR therules))
            (COND ( (compatible? arule theproblem)
                        (SETQ problembase
                                (APPEND (body (EVAL arule)) (CDR problembase)))
                        (writedecomp arule theproblem)
                        (RETURN therules)))
            (executeacycle theproblem (CDR therules))))

(DE effects (action) ;Called by executeacycle
    (PROG (modifs)
            (SETQ modifs (CASSQ action terminalactionstable))
            (COND (modifs (addon (CAR modifs)) (takeoff (CDR modifs))))))
```

```
(DE addon (listfacts) ;Called by effects
    (MAPC (LAMBDA (×)
            (COND ((NOT (MEMBER × factbase)) (NEWL factbase ×))))
        listfacts))

(DE takeoff (listfacts) ;Called by effects
    (MAPC (LAMBDA (×) (SETQ factbase (REMOVE × factbase))) listfacts))

(DE writeinsertion (action) ;Called by executeacycle
    (PRINT "action " action " inserted into the plan ===> fact base = "
        factbase))

(DE compatible? (therule theproblem) ;Called by executeacycle
    (PROG (patterns)
        (SETQ patterns (trigger (EVAL therule) NIL))
        (COND ( (OR (EQ '* (CAR patterns)) (EQUAL (CAR patterns)
                theproblem))
                (compfacts (CDR patterns ))))))

(DE trigger (therule result) ;Called by compatible? and trigger
    (COND ((EQ '<-- (CAR therule)) (REVERSE result))
          (T (trigger (CDR therule) (NEWL result (CAR therule))))))

(DE compfacts (factpatterns) ;Called by compatible?
    (COND ((NULL factpatterns))
          ((MEMBER (CAR factpatterns) factbase)
           (compfacts (CDR factpatterns)))))

(DE writedecomp (rule problem) ;Called by executeacycle
    (PRINT rule " decomposes " problem " ===> problem base = "
        problembase))

(DE body (therule) (CDR (MEMBER '<-- therule))) ;Called by executeacycle
```

Instantiation program 2.3

The instantiation program below is the directing coding of the symbolic knowledge base of Fig. 2.1.

```
(DF planlog theproblems
    ;Procedure for calling a symbolic expert system using engine 2.4
    (SETQ rulebase '(R1 R2 R3 R4 R5 R6 R7)
            R1 '(P1 <-- A1 A2)
            R2 '(P1 <-- A3)
            R3 '(P2 <-- P1 A4 P3)
            R4 '(P3 F1 <-- A5 P4)
            R5 '(P3 F2 <-- P5)
            R6 '(P3 <-- A4 P6)
            R7 '(P4 F2 F3 <-- A2)
            factbase '(F3 F4)
            terminalactionstable '((A1 (F1 F2)) (A2 () F1) (A3 (F1)) (A4) (A5
                        (F2) F4))
            problembase theproblems)
    (planretro))
```

Session 2.6

Session 2.6 below is obtained after the preceding instantiation program and engine-program 2.5 have both been loaded. On the first call to planlog the initial problem base is set to: (P1) while the second call to planlog corresponds exactly to the situation of Fig. 2.1 analysed in detail in Fig. 2.2 and 2.3 (note: the trace obtained from calling (planlog P2) is very close to Fig. 2.3).

(planlog P1)
r1 decomposes p1 ===> problem base = (a1 a2)
action a1 inserted into the plan ===> fact base = (f2 f1 f3 f4)
action a2 inserted into the plan ===> fact base = (f2 f3 f4)

PLAN FOR THE SERIES OF PROBLEMS SUBMITTED:
        a1
        a2
= success

(planlog P2)

r3 decomposes p2 ===> problem base = (p1 a4 p3)
r1 decomposes p1 ===> problem base = (a1 a2 a4 p3)
action a1 inserted into the plan ===> fact base = (f2 f1 f3 f4)
action a2 inserted into the plan ===> fact base = (f2 f3 f4)
action a4 inserted into the plan ===> fact base = (f2 f3 f4)
r5 decomposes p3 ===> problem base = (p5)

no (more) rules applicable to p5 ===> BACKTRACK
facts: (f2 f3 f4)
problems: (p3)
plan: (a1 a2 a4)

r6 decomposes p3 ===> problem base = (a4 p6)
action a4 inserted into the plan ===> fact base = (f2 f3 f4)

no (more) rules applicable to p6 ===> BACKTRACK
facts: (f2 f3 f4)
problems: (p3)
plan: (a1 a2 a4)

no (more) rules applicable to p3 ===> BACKTRACK
facts: (f3 f4)
problems: (p1 a4 p3)
plan: ()

r2 decomposes p1 ===> problem base = (a3 a4 p3)
action a3 inserted into the plan ===> fact base = (f1 f3 f4)
action a4 inserted into the plan ===> fact base = (f1 f3 f4)
r4 decomposes p3 ===> problem base = (a5 p4)
action a5 inserted into the plan ===> fact base = (f2 f1 f3)
r7 decomposes p4 ===> problem base = (a2)
action a2 inserted into the plan === fact base = (f2 f3)

PLAN FOR THE SERIES OF PROBLEMS SUBMITTED:
        a3

```
            a4
            a5
            a2
= success
```

2.3.4   Some simple improvements

Searching for all solutions

Several action plans that solve a given problem or series of problems may exist. Starting from the rules, initial facts and table of terminal actions in Fig. 2.1, 2 plans can be found to solve the problem P1: the plan (A1 A2) and the plan (A3). If an eighth rule, P1 <−− A1, were added, 2 plans solving P2 would be found: (A3 A4 A5 A2) and (A1 A4 A5 A2). Engine 2.5 can easily be adapted so that, at the user's request, it will successively search for the range of possible plans: solve must simply be made to fail in this case. The new ways of writing planretro and solve are presented below.

At the same time, small modifications have been introduced to allow the user to decide how much detail is given in the execution trace; a "degree of detail" with any value other than 0 or 1 will cause a detailed trace similar to that received in session 2.6; a degree of detail equal to 0 will give a trace which is reduced to just the plans built; a degree of detail equal to 1 will obtain an intermediate trace. Procedures writebacktrack and executeacycle have been altered slightly as a consequence.

The new sections are underlined.

```
(DE planretro () ;main procedure of the engine
    (PROG (planbase)
        (TERPRI) (PRINT "degree of detail (0: none, 1: a little, other: full)")
        (SETQ detail (READ)) (TERPRI)
        (COND ( (NULL (resolve problembase factbase planbase rulebase))
                (RETURN 'failure)))
        (RETURN 'success)))

(DE writeplan (steps) ;Called by planretro
    (MAPCAR (LAMBDA (x) (PRINCH '| | 12) (PRINT x)) steps))

(DE solve (problems facts plan series) ;Called by planretro and solve
    (PROG (cycleresult)
        (COND ( (NULL problembase)
                (TERPRI) (PRINT "A PLAN FOR THE SERIES OF PROBLEMS
                                SUBMITTED:")
                (writeplan (REVERSE planbase))
                (TERPRI) (PRINT "another? (yes or no)")
                (COND ((MEMBER (READ) '(yes o oui si T))
                        (TERPRI) (RETURN NIL))
                    (T (RETURN T)))))
        (SETQ cycleresult (executeacycle (CAR problembase) series))
        (COND ((NULL cycleresult) (RETURN NIL))
            ((solve problembase factbase planbase rulebase)
             (RETURN T))
            ((EQ cycleresult 'primitive) (RETURN NIL)))
        (SETQ problembase problems factbase facts planbase plan)
        (COND ((NEQ 0 detail) (writebacktrack)))
        (solve problems facts plan (CDR cycleresult)))))
```

```
(DE writebacktrack () ;Called by solve
    (PRINT "BACKTRACK")
    (COND ( (NEQ 1 detail)
              (PRINT "facts: " factbase)
              (PRINT "problems: " problembase)
              (PRINT "plan: " (REVERSE plan))))
    (TERPRI))

(DE executeacycle (theproblem therules) ;Called by solve and executeacycle
    (PROG (arule)
        (COND ( (ASSQ theproblem terminalactionstable)
              (NEXTL problembase)
              (NEWL planbase theproblem) (effects theproblem)
              (COND ((NEQ 0 detail) (writeinsertion theproblem)))
              (RETURN 'primitive))
            ( (NULL therules)
              (COND ((NEQ 0 detail) (TERPRI)
                    (PRIN "no (more) rules applicable to " theproblem
                          " ===> ")))
              (RETURN NIL)))
                    (SETQ arule (CAR therules))
                    (COND ( (compatible? arule theproblem)
                          (SETQ problembase
                    (APPEND (body (EVAL arule)) (CDR problembase)))
                    (COND ( (AND (NEQ 0 detail) (NEQ 1 detail))
                          (writedecomp arule theproblem)))
                    (RETURN therules)))
        (executeacycle theproblem (CDR therules))))
```

Session 2.7 below is obtained after joint loading of instantiation program 2.3 and engine 2.5 modified as indicated above. Two plans are indeed obtained for problem P1, but still just one for P2.

Session 2.7
(planlog P1)

degree of detail (0: none, 1: a little, other: full)
0

A PLAN FOR THE SERIES OF PROBLEMS SUBMITTED:
         a1
         a2

another? (yes or no)
yes

A PLAN FOR THE SERIES OF PROBLEMS SUBMITTED:
         a3

another? (yes or no)

yes

= failure

(planlog P2)

degree of detail (0: none, 1: a little, other: full)
0

A PLAN FOR THE SERIES OF PROBLEMS SUBMITTED:
>   a3
>   a4
>   a5
>   a5

another? (yes or no)
yes

= failure

Session 2.8 below is obtained after joint loading of instantiation program 2.3 with the addition of rule R8 (see **planlog** below) and engine 2.5 with the previously mentioned modifications. This time three plans are obtained for problem P1 and two for P2 (six for the series P1, P2). First of all, here is the result of introducing R1:

```
(DF planlog theproblems
    ;Procedure for calling a symbolic expert system using engine 2.4
    (SETQ rulebase '(R1 R2 R3 R4 R5 R6 R7)
          R1 '(P1 <−− A1 A2)
          R2 '(P1 <−− A3)
          R3 '(P2 <−− P1 A4 P3)
          R4 '(P3 F1 <−− A5 P4)
          R5 '(P3 F2 <−− P5)
          R6 '(P3 <−− A4 P6)
          R7 '(P4 F2 F3 <−− A2)
          R8 '(P1 <−− A1)
          factbase '(F3 F4)
          terminalactionstable '((A1 (F1 F2)) (A2 () F1) (A3 (F1)) (A4) (A5
                                  (F2) F4))
                                        problembase theproblems)
          (planretro))
```

Session 2.8
(planlog P1)

degree of detail (0: none, 1: a little, other: full)
1

action a1 inserted into the plan ===> fact base = (f2 f1 f3 f4)
action a2 inserted into the plan ===> fact base = (f2 f3 f4)

A PLAN FOR THE SERIES OF PROBLEMS SUBMITTED

                        a1
                        a2

another? (yes or no)
yes

BACKTRACK

action a3 inserted into the plan ===> fact base = (f1 f3 f4)

A PLAN FOR THE SERIES OF PROBLEMS SUBMITTED:
                        a3

another? (yes or no)
yes

BACKTRACK

action a1 inserted into the plan ===> fact base = (f2 f1 f3 f4)

A PLAN FOR THE SERIES OF PROBLEMS SUBMITTED
                        a1

another? (yes or no)
yes

BACKTRACK

no (more) rules applicable to p1 ===> = failure

(planlog P2)

degree of detail (0: none, 1: a little, other: full)
0

A PLAN FOR THE SERIES OF PROBLEMS SUBMITTED:
                        a3
                        a4
                        a5
                        a2

another? (yes or no)
yes

A PLAN FOR THE SERIES OF PROBLEMS SUBMITTED:
                        a1
                        a4
                        a5
                        a2

another? (yes or no)
yes

= failure

(planlog P1)

degree of detail (0: none, 1: a little, other: full)
0

A PLAN FOR THE SERIES OF PROBLEMS SUBMITTED:
                a1
                a2

another? (yes or no)
no
= success

(planlog P3)

degree of detail (0: none, 1: a little, other: full)
1

action a4 inserted into the plan ===> fact base = (f3 f4)

no (more) rules applicable to p6 ===> BACKTRACK

no (more) rules applicable to p3 ===> = failure

(planlog P1 P2)

degree of detail (0: none, 1: a little, other: full)
0

A PLAN FOR THE SERIES OF PROBLEMS SUBMITTED:
                a1
                a2
                a3
                a4
                a5
                a2

another? (yes or no)
yes

A PLAN FOR THE SERIES OF PROBLEMS SUBMITTED:
                a1
                a2
                a1
                a4
                a5
                a2

another? (yes or no)
yes

A PLAN FOR THE SERIES OF PROBLEMS SUBMITTED:
                a3
                a3
                a4
                a5
                a2

another? (yes or no)
yes

A PLAN FOR THE SERIES OF PROBLEMS SUBMITTED:

                        a3
                        a1
                        a4
                        a5
                        a2

another? (yes or no)
yes

A PLAN FOR THE SERIES OF PROBLEMS SUBMITTED:
                        a1
                        a3
                        a4
                        a5
                        a2

another? (yes or no)
yes

A PLAN FOR THE SERIES OF PROBLEMS SUBMITTED:
                        a1
                        a1
                        a4
                        a5
                        a2

another? (yes or no)
yes

= failure

another? (yes or no)
yes

A PLAN FOR THE SERIES OF PROBLEMS SUBMITTED:
                        a1
                        a3
                        a4
                        a5
                        a2

another? (yes or no)
yes

A PLAN FOR THE SERIES OF PROBLEMS SUBMITTED:
                        a1
                        a1
                        a4
                        a5
                        a2

another? (yes or no)
yes

= failure

Interaction

It would not be difficult to borrow the characteristics of engines 2.2, 2.3 and 2.4 and allow planning engine 2.5 to function interactively. For example, when no rule can be applied to the head of problembase the user could be asked whether a backtrack should be carried out; the head of problembase and all or part of factbase could then be displayed; the user might perhaps say that the problem given to him can be solved under certain conditions.

Enhancing the effects of terminal actions

The ways in which facts were interpreted and the terminal actions table introduced at the beginning of section 2.3 was used were marred by ambiguity: when the effects of an action involve the addition of a fact, were we or were we not to ensure that the number of times this fact occurred in the fact base was not increased? In practice, the ambiguity was removed by the definition of the procedures effects, addon and takeoff: in particular, a fact can occur only once in factbase. For certain applications it could be desirable to have several occurrences of a single fact. More generally, the effects of the terminal actions could be far broader than the simple addition or suppression of expressions which are totally instantiated at the time when the terminal actions table is written. For instance, inserting the action (increment) into the plan might lead to trying to find a list-fact of the form (counter $<n>$) — perhaps (counter 5) at the current moment — and replacing the integer $<n>$ by the following integer, ($<n> + 1$). To allow such extensions it is necessary to redefine (i) the structure of terminalactionstable and (ii) the procedures effects, addon and takeoff. Note that enhancing the possible effects of the terminal actions is equivalent to enhancing the interpretation of the *action* or *body* part of the rules.

Mechanisms to prevent looping and end of the planning

Breaking down a problem P by means of a rule R can lead to breaking down the problem P by another rule R′ which in turn may use R. If, when R is called for a second time, the fact base is in exactly the same state as when R was first called, then looping is taking place. To detect looping, whenever any attempt is made to decompose a problem P by means of a rule R we can test whether this second decomposition *is a descendant of* the first (that is to say whether, in the AND/OR graph representing the problems by nodes and the decompositions performed by hyperarcs between each father and its children, the second occurrence of P is represented by a node which is a descendant of the one representing the first) and whether the state of the fact base is the same at the time of the 2 decompositions (same *context*). We must be careful: if the second decomposition of P by R is called in the same context as the first but the second decomposition does not come from the first, then there is no looping.

The fact base manipulated by engine 2.5 can only take a *finite* number of distinct states since (i) all expressions of possible facts appear in one or other part of the knowledge base (in the rules, facts or terminal actions table) from the outset (hence forming a finite set) and (ii) the expression of any fact never occurs more than once in the fact base (hence a finite set of combinations). The set of rules is hypothesised to always be finite. Consider the AND/OR graph which can be obtained by applying the rules in all possible ways but forbidding the application of a rule in the same context to two problems where one is a descendant of the other; this graph (let us call it the *proper graph of the sub-problems*) is finite. Consequently, if the mechanism for detecting loops suggested in the previous paragraph is implemented, we gain the theoretical assurance that the planning process will terminate (even though it may prove cumbersome) on finding a *solution-plan* if one exists (a plan which solves the series of problems submitted).

This will not be the case if the effects of the terminal actions are of such a nature that the fact base can take an *infinite* number of distinct states (see the section about enhancing the effects of terminal actions): in this situation the proper graph of the sub-problems is infinite. In order to avoid looping, the ancestors, of which there are a finite but unspecified number, of every problem which will be decomposed must be examined, which can be very costly. Despite these tests for detecting loops, the depth first strategy can not guarantee that the engine will stop even if a solution-plan exists.

Controlling the depth of development

Engine-program 2.6 below is a new version of engine 2.5. It controls the depth first chaining of problem decompositions. The user states a depth which is not to be exceeded. The engine is thus guaranteed to halt.

This kind of control is especially desirable if the engine is intended to run on a fact base that can take an infinite number of distinct states (see the two preceding paragraphs concerning (i) the effects of terminal actions and (ii) stopping the engine). It is useful too when the fact base can only take a finite number of states but there is no control over whether a rule is applied in the same context to two problems, one of which is a descendant of the other.

In both cases there may be some reason to believe that exceeding a certain level of decomposition is pointless; this type of control can then be useful to limit the depth of the search so as not to expend too much effort.

This new version of engine 2.5 incorporates the modifications described at the start of §2.3.4 which deal with searching for all solutions. In addition, the passing of parameters to executeacycle, compatible? and writedecomp has been slightly simplified: in all 3 cases the current problem is now taken directly from the head of problembase.

planretro asks the user what depth-limit is desired. A *depth* equal to 0 is associated with each initial element of problembase; elements of problembase appearing through decomposition of a problem are assigned a *depth* equal to that of the decomposed problem, plus 1; the procedure givedepth, called by planretro and decompose, gathers every problem expression preceded by its depth into a list; the new structure of the elements in problembase entails a few minor alterations to executeacycle (line 3), compatible? and writedecomp. The depth-limit and the depth of the current problem are compared in executeacycle; note that when the limit is reached the engine forces a backtrack exactly as if no (more) rules applicable to the problem existed.

Engine-program 2.6

Modifications to procedure in engine-program 2.5 and names of new procedures are underlined.

```
(DE planretro () ;main procedure of the engine
    (PROG (planbase)
        (SETQ problembase (givedepth 0 problembase))
        (TERPRI)
        (PRINT "degree of detail (0: none, 1: a little, other: full)")
        (SETQ detail (READ))
        (TERPRI)
        (PRINT "depth limit ?")
        (SETQ depthlimit (READ))
        (TERPRI)
        (COND ( (NULL (resolve problembase factbase planbase rulebase))
                (RETURN 'failure)))
        (RETURN 'success)))

(DE writeplan (steps) ;Called by planretro
    (MAPCAR (LAMBDA (×) (PRINCH '| | 12) (PRINT ×)) steps))

(DE solve (problems facts plan series) ;Called by planretro and solve
    (PROG (cycleresult)
        (COND ( (NULL problembase)
                (TERPRI) (PRINT "A PLAN FOR THE SERIES OF PROBLEMS
                    SUBMITTED:")
                (writeplan (REVERSE planbase))
                (TERPRI) (PRINT "another? (yes or no)")
                (COND ((MEMBER (READ) '(yes o oui si T))
                        (TERPRI) (RETURN NIL))
            (T (RETURN T)))))
        (SETQ cycleresult (executeacycle series))
        (COND ((NULL cycleresult) (RETURN NIL))
                ((solve problembase factbase planbase rulebase)
                 (RETURN T))
                ((EQ cycleresult 'primitive) (RETURN NIL)))
        (SETQ problembase problems factbase facts planbase plan)
        (COND ((NEQ 0 detail) (writebacktrack)))
        (solve problems facts plan (CDR cycleresult))))

(DE writebacktrack () ;Called by solve
    (PRINT "BACKTRACK")
    (COND ( (NEQ 1 detail)
            (PRINT "facts: " factbase)
            (PRINT "problems: " problembase)
            (PRINT "plan: " (REVERSE plan))))
    (TERPRI))
```

```
(DE executeacycle (therules) ;Called by solve and executeacycle
    (PROG (theproblem currentdepth arule)
        (SETQ theproblem (CADAR problembase) currentdepth (CAAR
                                        problembase))
        (COND ( (ASSQ theproblem terminalactionstable)
                (NEXTL problembase) (NEWL planbase theproblem) (effects
                theproblem)
                (COND ((NEQ 0 detail) (writeinsertion theproblem)))
                (RETURN 'primitive))
              ( (NULL therules)
                (COND ((NEQ 0 detail) (TERPRI)
                    (PRIN "no (more) rules applicable to " theproblem
                        " ===> ")))
                (RETURN NIL))
              ( (= currentdepth depthlimit)
                (COND ( (NEQ 0 detail) (TERPRI)
                    (PRIN "depth limit reached on " theproblem
                        " ===> ")))
                (RETURN NIL))
        (SETQ arule (CAR therules))
        (COND ( (compatible? arule)
                (decompose arule currentdepth)
                (RETURN therules)))
        (executeacycle theproblem (CDR therules))))

(DE decompose (therule currentdepth) ;Called by executeacycle
    (COND ((AND (NEQ 0 detail) (NEQ 1 detail)) (writedecomp arule)))
    (SETQ problembase
        (APPEND (givedepth (1+ currentdepth) (body (EVAL arule)))
                (CDR problembase))))

(DE givedepth (childdepth childproblems) ;Called by planretro and decompose
    (MAPCAR (LAMBDA (x) (LIST childdepth x)) childproblems))

(DE effects (action) ;Called by executeacycle
    (PROG (modifs)
        (SETQ modifs (CASSQ action terminalactionstable))
        (COND (modifs (addon (CAR modifs)) (takeoff (CDR modifs))))))

(DE addon (listfacts) ;Called by effects
    (MAPC (LAMBDA (x) (COND ((NOT (MEMBER x factbase))
            (NEWL factbase x))))
        listfacts))

(DE takeoff (listfacts) ;Called by effects
    (MAPC (LAMBDA (x) (SETQ factbase (REMOVE x factbase))) listfacts))

(DE writeinsertion (action) ;Called by executeacycle
    (PRINT "action " action " inserted into the plan ===> fact base = "factbase))

(DE compatible? (therule) ;Called by executeacycle
    (PROG (patterns)
        (SETQ patterns (trigger (EVAL therule) NIL))
        (COND ( (OR (EQ '* (CAR patterns))
                (EQUAL (CAR patterns) (CADAR problembase)))
                (compfacts (CDR patterns ))))))
```

```
(DE trigger (therule result) ;Called by compatible? and trigger
    (COND ((EQ '<-- (CAR therule)) (REVERSE result))
               (T (trigger (CDR therule) (NEWL result (CAR therule)))))))

(DE compfacts (factpatterns) ;Called by compatible?
    (COND ((NULL factpatterns))
               ((MEMBER (CAR factpatterns) factbase)
                (compfacts (CDR factpatterns)))))

(DE writedecomp (rule) ;Called by executeacycle
    (PRINT rule " decomposes " (CADAR problembase)
               " ===> problem base = " problembase))

(DE body (therule) (CDR (MEMBER '<-- therule))) ;Called by executeacycle
```

Session 2.9

The session shown below is obtained after both engine-program 2.6 and instantiation program 2.3 have been loaded. When planlog is called for the second time, the depth-limit imposed is too shallow for it to be able to construct a plan.

(planlog P2)

degree of detail (0: none, 1: a little, other: full)
2

depth limit ?
3

r3 decomposes p2 ===> problem base = ((1 p1) (1 a4) (1 p3))
r1 decomposes p1 ===> problem base = ((2 a1) (2 a2) (1 a4) (1 p3))
action a1 inserted into the plan ===> fact base = (f2 f1 f3 f4)
action a2 inserted into the plan ===> fact base = (f2 f3 f4)
action a4 inserted into the plan ===> fact base = (f2 f3 f4)
r5 decomposes p3 ===> problem base = ((2 p5))

no (more) rules applicable to p5 ===> BACKTRACK
facts: (f2 f3 f4)
problems: ((1 p3))
plan: (a1 a2 a4)

r6 decomposes p3 ===> problem base = ((2 a4) (2 p6))
action a4 inserted into the plan ===> fact base = (f2 f3 f4)

no (more) rules applicable to p6 ===> BACKTRACK
facts: (f2 f3 f4)
problems: ((1 p3))
plan: (a1 a2 a4)

no (more) rules applicable to p3 ===> BACKTRACK
facts: (f3 f4)
problems: ((1 p1) (1 a4) (1 p3))
plan: ()

r2 decomposes p1 ===> problem base = ((2 a3) (1 a4) (1 p3))
action a3 inserted into the plan ===> fact base = (f1 f3 f4)
action a4 inserted into the plan ===> fact base = (f1 f3 f4)
r4 decomposes p3 ===> problem base = ((2 a5) (2 p4))
action a5 inserted into the plan ===> fact base = (f2 f1 f3)
r7 decomposes p4 ===> problem base = ((3 a2))
action a2 inserted into the plan ===> fact base = (f2 f3)

A PLAN FOR THE SERIES OF PROBLEMS SUBMITTED:

> a3
> a4
> a5
> a2

another? (yes or no)
no
= success

(planlog P2)

degree of detail (0: none, 1: a little, other: full)
1

depth limit ?
2

action a1 inserted into the plan ===> fact base = (f2 f1 f3 f4)
action a2 inserted into the plan ===> fact base = (f2 f3 f4)
action a4 inserted into the plan ===> fact base = (f2 f3 f4)

depth limit reached on p5 ===> BACKTRACK

action a4 inserted into the plan ===> fact base = (f2 f3 f4)

depth limit reached on p6 ===> BACKTRACK

no (more) rules applicable to p3 ===> BACKTRACK

action a3 inserted into the plan ===> fact base = (f1 f3 f4)
action a4 inserted into the plan ===> fact base = (f1 f3 f4)
action a5 inserted into the plan ===> fact base = (f2 f1 f3)

depth limit reached on p4 ===> BACKTRACK

action a4 inserted into the plan ===> fact base = (f1 f3 f4)

depth limit reached on p6 ===> BACKTRACK

no (more) rules applicable to p3 ===> BACKTRACK

no (more) rules applicable to p1 ===> BACKTRACK

no (more) rules applicable to p2 ===> = failure

Ability to pose additional problems
In session 2.8 the call: (planlog P3) failed; in fact, rule R6 is the only one
that can be triggered and doing so leads to a dead-end. In such a case it would

be desirable for the system to try to remove the obstacles to the triggering of currently untriggerable rules such as R4. Here, for example, F1 would have to be established; terminal actions A1 and A3 are capable of doing this; now, A1 could be executed if the resolution of P1 by R1 were invoked, while A3 could be executed if R2 were invoked to solve P1; R2 is preferable to R1 since action A2 eliminates F1 established by A1. So, in order to solve P1, it would be a good idea for the system to *set itself* the additional problem P1 and then attack it via R2; the engine could produce the plan: (A3, A5, A2) by this route.

Engines 2.5 and 2.6 do not make use of all the information represented by the knowledge base in its entirety. They could be easily modified so that:

— at least when necessary, a rule like R4 would be treated as though it were written: P3 <−− F1, A5, P4 (which would cause rules to be sought that solve (establish) F1), and
— at least when necessary, rules like: F1 <−− A1 or F1 <−− A3 would be extracted from the table of terminal actions.

Short of these changes, a rudimentary manner of overcoming the failure of engines 2.5 and 2.6 is to modify instantiation program 2.3 as follows:

```
(DF planlog theproblems
    (SETQ rulebase '(R1 R2 R3 R4 R40 R5 R50 R6 R7 R70 R8 R9 R10 R11)
              R1 '(P1 <−− A1 A2)
              R2 '(P1 <−− A3)
              R3 '(P2 <−− P1 A4 P3)
              R4 '(P3 F1 <−− A5 P4)
              R40 '(P3 <−− F1 A5 P4)
              R5 '(P3 F2 <−− P5)
              R50 '(P3 <−− F2 P5)
              R6 '(P3 <−− A4 P6)
              R7 '(P4 F2 F3 <−− A2)
              R70 '(P4 <−− F2 F3 A2)
              R8 '(F1 <−− A1)
              R9 '(F2 <−− A1)
              R10 '(F1 <−− A3)
              R11 '(F2 <−− A5)
              factbase '(F3 F4)
              terminalactionstable '( (A1 (F1 F2)) (A2 () F1) (A3 (F1))
                                      (A4)(A5 (F2) F4))
              problembase theproblems)
    (planretro))
```

Two plans are then obtained: (A1 A5 A2) and (A3 A5 A2).

2.3.5   Other desirable capacities for planning systems

We now point out two sorts of inadequacy in engines 2.5 and 2.6 which would require more delicate alterations than those suggested above.

Capacity to deal with unordered problems and to investigate unordered decomposition rules

Engines 2.5 and 2.6 can only accept *ordered conjunctions* of problems: problems to be resolved one after the other *in the order in which the conjunction is expressed*, with the effects of each resolution being passed on to the next. But *unordered* conjunctions of problems cannot be dealt with. Thus, consider the unordered conjunction: {P1 P2}. According to whether this is treated as one or other of the ordered conjunctions (P1, P2) or (P2, P1) a solution-plan may or may not be found. To solve an unordered conjunction of problems it would either be necessary to integrate into the previous engines a general mechanism capable of anticipating, when necessary, the various permutations of a collection of problems, or else equipping the engines with the ability to interpret the user's rules acting on the problem base and supplying them with specific permutation rules for the particular collection of problems considered.

The difficulty grows, as does the number of potential solution-plans, if the knowledge base contains rules for breaking problems down into *unordered* sub-problems. Take for example the rules: P1 <−− {P3, P4} and P2 <−− {P5, P6} and the unordered conjunction {P1, P2}; when decomposition takes place it may be just as relevant to consider the intermixed series (P3, P5, P4, P6) or (P6, P4, P3, P5) as the juxtaposed series (P5, P6, P4, P3) or (P3, P4, P6, P5).

Capacity to "keep" already completed solutions

Let us consider the problem represented by Fig. 2.4:

Fig. 2.4.

This problem could be represented by: given "A ON FLOOR", "C ON A" and "ON FLOOR", establish "A ON B" and "B ON C" using terminal actions of the form: "TAKE x" and "PUT x ON y" (for which the types of arguments x and y and the preconditions for use would have to be specified). Depending on whether "A ON B" is solved first and then "B ON C" or the opposite order is chosen, each sub-problem is solved but the final situation is not the one expected.

The same sort of difficulties as those seen in Fig. 2.4 can also be met starting from the knowledge base of Fig. 2.1. R1 could be interpreted as: "in order to solve P1, just execute action A1 and then action A2, so that F2 is obtained in the end". Then add rule R8: P5 $<--$ A3 and suppose that we wish to solve P5 which, following the same interpretation, means: obtain F1. If we try to solve P5 and P1, P5 will be solved by rule R8 and then P1 by rule R1. But by the end of this only F2 will still be established (apart from F3 and F4). P5 was solved "in passing" but is no longer so when P1 is dealt with (in the sense that F1 is no longer established). Note that if P1 is attacked before P5, both F1 and F2 remain established when the resolution is over.

Handling the problem base as a stack (*last in, first out*) is not sufficient for tackling these kinds of difficulties. For a more detailed examination of the difficulties and basic techniques involved in action planning, refer to [NILSSON 1980] and [SACERDOTI 1977].

2.4   FOR GENERATORS WITH VARIABLES: PATTERN-MATCHING

2.4.1   Variables in the rules

From §2.1.2 up until now, we have only considered knowledge representation languages which do not use *variables*. Inference engines that interpret a language for expressing rules and facts without allowing variables to be used are commonly called *0 engines*. Inference engines that interpret rules with variables (though not necessarily facts with variables) are called *1 engines*. This terminology stems from the similarities that can be drawn, appropriately or not, between the language interpreted by an engine and either the *language of propositional logic*, which does not contain variables (sometimes called **zero** order language or logic), or else the *language of* **first** *order predicate logic* which contains universally or existentially quantified variables. In the same way we can talk about *0 generators* and *1 generators* (beware: not everybody attributes the same meaning to these terms).

In 1 generators, a rule of the (symbolic) form: *premise*$(x, y, \ldots)$ $--->$ *conclusion*$(x, y, \ldots)$ is most frequently interpreted as: $\forall$x,y,... IF premise(x,y,...) is established THEN *conclusion*$(x, y, \ldots)$ is established. Hence forward-chaining will derive conclusion(x, y, . . .) for any x, y, . . . which establish *premise*$(x,y, \ldots)$; using backward-chaining, if the goal is to establish conclusion(x, y, . . .) for certain fixed values of x, y, . . ., the new goal *premise*$(x,y, \ldots)$ for the same values of x, y, . . . can be considered. In the majority of 1 generators, facts can not include variables. For example, if we have the *established fact*: *premise*$(a, b, \ldots)$ in a forward-chaining engine, where a and b are constants, we can derive the new established fact: *conclusion*$(a, b, \ldots)$; if we have the *fact to be established* (hence goal or problem): *conclusion*$(a, b, \ldots)$, a backward-chaining engine can derive the new fact to be established: *premise*$(a, b, \ldots)$. Speaking generally, the operation which allows us to recognise that a certain fact (which may or may not be established, depending on the generators) is *compatible* with a certain element in the target of a rule and that *pairs* (or *matches*) the rule's variables with sub-expressions of the fact is called *pattern-matching*. By extension, this term also refers to one of the 4 major steps in the basic cycle of inference engines: it is the one which, by comparing a set of rules and a set of facts, allows a subset of rules to be discovered whose firers are compatible with the state of the collection of facts.

The use of variables in a rule allows several rules (or even an infinite number) to be condensed into a single one, which is known as *factorising* knowledge.

Examples of generators permitting variables in the rules (sometimes in a restricted fashion) but not in the facts are: EMYCIN [VAN MELLE 1979] [BUCHANAN and SHORTLIFFE 1984] and its numerous derivatives (notably M1 and S1 developed by the Ste Framentec, PC Consultant developed by Texas Instruments and TG1 developed by Cognitech), TRO-PIC [LATOMBE 1977], OPS5 [FORGY 1980] [BROWNSTON 1985], ARGOS-II [FARRENY 1980] [PICARDAT 1985], SNARK [LAURIERE 1986], TANGO [CORDIER and ROUSSET 1982] [ROUSSET 1983] and GURU (distributed by the society ISE-CEGOS).

Fig. 2.4A gives the text of a rule in ARGOS-II. After its name, r9, there is a trigger made up of 3 elementary *patterns*: (take >x), (<x room >y *) and ABSENT (<y * radioactive state *)

```
(r9 ( (take >x) (<x room >y *) ABSENT (<y radioactive state *))
     ( (EXECUTE (point-in-room (<y) (z) ))
       (REPLACEPB (go <z) (approach <x))
       (RECORD (PLAN (grasp (<x) ))
       (UPDATE (MODIFY (robot *) (taken <x)) ))))
```

Fig. 2.4A.

In order to decide whether to trigger rule r9, the first pattern, (take>x), will be compared with the various elements in the *base of facts to be established*. A rule can only be fired if all of its patterns have been found to be compatible with the fact base under consideration. If there is a fact to be established such as: (take test-tube), the pattern will be matched and the variable x (> signifies a variable *to be assigned*) will be assigned the value test-tube. The following patterns (if any) refer to the *base of established facts*. The notation <x represents the last value assigned to variable x (< signifies a variable *to be read*). This means that the pattern (<x room >y *) is now equivalent to the pattern (test-tube room >y *). Suppose the established fact: (test-tube room r4 table t3) exists. r4 will be assigned to variable y; the *pattern-matching operator* * is defined to be compatible with any (possibly empty) sequence of expressions. The pattern: ABSENT (<y * radioactive state *) is now equivalent to the pattern: ABSENT (r4 * radioactive state *). As a result of the role played by the pattern-matching operator ABSENT, the pattern will not be considered to be compatible with the fact base unless there is no fact in the latter which is compatible with the sub-pattern: (r4 * radioactive state *).

We should point out that the rule's body contains a sequence of 3 principal *directives*: EXECUTE, REPLACEPB and RECORD.

Note that in first order predicate logic rules and facts are not distinguished a priori. The language PROLOG only knows about *Horn clause* type formulae in first order predicate logic, which are disjunctions of literals of which at most one is positive; in this special case it is customary to treat clauses reduced to a single positive literal as established facts, clauses containing only negative literals as facts to be refuted (meaning they are to be proved false), or goals, and lastly all other clauses as rules. In the PROLOG inference engine, pattern-matching tries to match a goal with either an established fact or the positive literal of a rule (a rule's trigger is its positive literal); a notable feature of PROLOG is that both arguments in pattern-matching, namely the goal **and** the rule, may contain variables. In chapter 3 we shall return to PROLOG in detail and examine the mode of pattern-matching associated with it, called *unification*.

In the rest of this chapter, we shall assume that only one of the two arguments in pattern-matching (the one representing the trigger of a rule or an element in the trigger) can contain variables. This kind of — asymmetric — pattern-matching is often called *semi-unification*. Nonetheless, semi-unification can not in general be classed as a particular instance of classical unification (see [FARRENY & GHALLAB 1986]).

2.4.2 The roles of pattern-matching

Expert systems and expert system generators were originally referred to as Pattern Directed Inference Systems (PDIS) [WATERMAN AND HAYES ROTH 1978]. The merit of this term was that it stressed the central role played by pattern-matching in the *invocation* of rules (via their triggers) in that type of system. Pattern-matching can however be used in circumstances other than invocation. It can be used in the body part of rules to *seek information* in the knowledge bases available (rule bases, fact bases or other sorts). It can also be used to break down an information structure into sub-structures and to build new structures.

Pattern matching for invoking rules

The use of pattern-matching in **writing rule triggers** allows the conception of the generator (including the rule and fact language) to be separated from the construction of knowledge bases which depend on applications and their evolution. When someone introduces a fact into the fact base, he does not have to know (to variable degrees, according to the circumstances) about the identity, form, nature or even existence of rules whose triggers involve patterns which are compatible with the fact. Conversely, when someone introduces a new rule or retracts an existing one, he does not have to know what relationships to the available facts and other rules are thereby created or broken.

In ARGOS-II, for example (see Fig. 2.4), when the engine decides to tackle the problem (or fact to be established): (take test-tube), it uses pattern-matching comparing **this** *problem instance* with the **whole** of the rule base (in actual fact the set of *unsuppressed* rules) to determine which rules are capable of reducing the problem, which are those whose first pattern in the trigger part is compatible with (take test-tube). The engine then carries out an *associative access of the rule base*: any rule can be selected starting from a fragmentary description (the problem instance) of its contents, and this description may be variable (different problem instances). The fact that, for **each** sub-problem to be solved, *all* the rules are again compared is costly in execution time. When a rule has been selected by matching its first pattern and (take test-tube), the engine compares any further patterns in the rule with the base of established facts; one after another, it performs a pattern-match between **each** pattern and the **whole** of the base of established facts; the engine then carries out an *associative access of the established fact base*. In order to reduce the execution costs, *database compilation* techniques have been suggested (see in particular [GHALLAB and DUFRESNE 1982, 1984]). However, whether or not the engine compiles the knowledge, it is important for the user who has to construct or revise the knowledge base to be able to count on *associative access* (of the rules or facts) in order to express the rule triggers in a flexible manner without having complete knowledge of the state of the database at the moment when these triggers are examined by the engine.

Pattern-matching for associative access of knowledge

Allowing pattern-matching to be used **in rule bodies**, so as to access the different parts of the knowledge base associatively, strengthens the flexibility and power of the language for encoding knowledge. The use of pattern-matching permits knowledge to be represented flexibly (an identifier does not have to be chosen and known for each item of knowledge) and adaptably (according to the features currently of interest). In Fig. 2.4, for example, the expression MODIFY (robot *) causes the base of established facts to be scanned for a list beginning with the word robot, the remainder of the list being irrelevant; the same fact can be the target of different patterns, and several distinct facts can be the target of the same pattern.

Pattern-matching for decomposing/reconstructing information

Consider the expression (MODIFY (robot *) (taken <x)) in the rule in Fig. 2.4. Suppose that test-tube has been assigned to x on invoking the rule and suppose that the first established fact compatible with (robot *) is (robot position p12). The MODIFY primitive causes (robot position p12) to be replaced by (robot position p12 taken test-tube) in the storage of established facts. The presence of an advanced language for writing patterns will allow pattern-matching to be used all the more effectively for decomposing and reconstructing information.

2.4.3   Pattern-matching as a data processing procedure

Pattern-matching is principally an operation *returning a compatibility value* between two objects. As stated earlier, this chapter only deals with the (highly representative) case where one of the objects is an instance (hereafter denoted I), meaning an expression formed from constants alone, while the other is a pattern (hereafter denoted P) allowing constants, variables and other symbols. Most expert systems consider the result of the operation to have a binary domain: either P and I are compatible, or else P and I are incompatible.

From here on we shall only take this classical, binary view of pattern-matching, although in a number of cases it would be particularly appropriate to permit a discrete or continuous gradation of compatibility between P and I (see [CAYROL 1980, 1882], [DUBOIS 1985]). Thus we shall term *pattern-matching procedure* a data processing procedure whose arguments are a pattern P and an instance I and which returns T or NIL according as P and I are or are not compatible. The way in which P and I are determined to be or not to be compatible characterises the *pattern-matching* mode considered; for example, when the pattern-matching mode considered is "literal identity", the pattern P = (five six) is judged to be incompatible with the instance I = (cinq six); on the other hand, in a pattern-matching mode where the constant five is taken to be compatible with the constant cinq, P and I could be judged compatible. A pattern-matching procedure implements a pattern-matching mode. The combination of a language for patterns and the associated pattern-matching procedure constitutes a *pattern-matching system*.

Apart from returning a compatibility value, the execution of a pattern-matching procedure can be accompanied by *side effects*. These might be assigning values to variable identifiers, or invoking procedures.

In practice, P and I are character strings or S-expressions (atoms and lists). Pattern-matching was used very early on in software for natural language understanding, shape recognition and text processing. To the best of our knowledge, the first programming language with pattern-matching primitives was SNOBOL (see [GRISWOLD 1971]). Independently of the PDIS, which by their nature all use pattern-matching, numerous languages developed in Artificial Intelligence work offer pattern-matching primitives. Examples are PLANNER, CONNIVER, QA4, SAIL, LISP70, FUZZY, QLISP, KRL, TELOS, AMORD and PLASMA. Depending on the individual author, the expression *pattern-matching* can be understood in a variety of alternative ways: *pairing, model recognition, pattern correspondence, coincidence of forms.* Similarly, *patterns* can be understood as *models, forms, profiles,* outlines.

To give a concrete idea of the languages for patterns available in certain generators and of how patterns may be interpreted, we now present, in 3 stages, a set of pattern-matching primitives inspired by the ones defined in ARGOS-II [FARRENY 1980] [PICARDAT 1985] and the corresponding pattern-matching procedure. Technical discussion of pattern-matching is also to be found in [WINSTON and HORN 1981] and [WERTZ 1985].

2.4.4 A pattern-matching system. First stage

First let us specify the vocabulary to be used. We shall call any atom (in the LISP sense) not starting with > or < a *constant*, any atom starting with > a *variable to be assigned*, any atom starting with < a *variable to be read*, and a variable to be assigned or read simply *variable*. The first character in a variable will be called the *prefix* and the rest will be the *radical*.

Pattern language

An *instance* I will be either a constant or some list of constants. A *pattern* will be either a constant, a variable or a list of constants and variables. A *well formed pattern* will be a pattern in which every occurrence of a variable to be read is preceded in the pattern (somewhere to the left in its expression) by a variable to be assigned bearing the same radical.

For example: (c1 >var1 (<var1 c2 >var3) c3 <var3 <var1 >var4 (c4) >var1) is a well formed pattern, while (c1 >var1 (<var1 c2 <var3) c3) is not. Let us now define a pattern-matching mode.

Pattern-matching mode

(a) If the pattern P is a constant, it is only compatible with the instance I if P and I are identical (hence, P = five is not compatible with I = cinq).

(b) If P is a variable to be assigned, P is compatible with any instance I; furthermore, the pattern-matching procedure associates I to P's radical as P's *value to be read*.

(c). If P is a variable to be read, P is only compatible with I if (i) P's radical already possesses a value to be read and (ii) this value is identical to I.

(d) If P is a list, P will only be compatible with I if (i) I is a list, (ii) the pattern defined by the CAR of P is compatible with the instance defined by the CAR of I and (iii) the pattern defined by the CDR of P is compatible with the instance defined by the CDR of I.

It is clear that if a pattern-matching procedure is invoked with a well formed initial argument P, any variable to be read in P will receive a value to be read before undergoing test c.

Session 2.10 below illustrates the pattern-matching mode. The pattern-matching procedure named pattern is defined further on (program 2.7); the value of the first argument to pattern is the pattern P and the value of the second argument is the instance I.

Session 2.10

```
(pattern '(surname >x first-name >y date-of-birth >z)
         '(surname Smith first-name John date-of-birth 29/2/44))
= t
x y z
= smith
= john
= 29/2/44

(pattern '(make ford model fiesta >x registration B 723 VOD owners >y >z)
         '(make ford model fiesta blue registration B 723 VOD owners Bru Jo))
= t
x y z
= blue
= bru
= jo

(pattern '(>x (<x >x >y) <x <y) '(a (a b c) b c))
= t
```

x y
= b
= c

Program 2.7
(DE **pattern** (p i) ; p: pattern, i: instance
    (COND ((ATOMP p)
         (COND ((EQ p i))
             ((EQ (caratom p) '>) (SET (cdratom p) i))
             ((EQ (caratom p) '<) (EQUAL (EVAL (cdratom p)) i))))
        ((ATOMP i) NIL)
        (T (AND (pattern (CAR p) (CAR i)) (pattern (CDR p) (CDR i))))))

DE **caratom** (atom) ; extracts the prefix of the atom argument
    (CAR (EXPLODECH atom)))

(DE **cdratom** (atom) ; extracts the radical of the atom argument
    (IMPLODECH (CDR (EXPLODECH atom))))

This follows the definition of the pattern-matching mode closely.

In this implementation, variables are atoms; they are detected by systematically testing for the presence of < or > as the prefix of all the atoms in the pattern. Hence the atoms have to be exploded into characters (by EXPLODECH); characters also have to be fused into atoms (by IMPLO-DECH) to recover variables' radicals. This processing may prove cumbersome if the same pattern is frequently used as an argument to pattern, in which case it is better to transform the pattern before any use of it, every variable of type >x or <x being replaced by a list (> x) or (< x); this solution is adopted and described in §3.2.4.3. It has also been decided to **assign** a variable's value to be read to the LISP atom which is the radical of the variable concerned; alternatively, an *association list* (or *a-list*) could be constructed to link variables' radicals and corresponding values to be read (for example, even numbered elements for one and odd numbered for the other).

2.4.5 A pattern-matching system second stage
The pattern-matching system introduced in the previous section is extended by first enhancing the pattern language and then clarifying the pattern-matching mode.

Language for patterns
Patterns can now be constants, variables, or lists of constants, variables and *pattern-matching operators*. Four kinds of pattern-matching operator are defined: *, ?, *n and ?n, where n stands for a whole number. Example patterns are:

    (>x ? >y), (>x ?2 >y), (>x * >y), (>x ?3 >y *), (a (>x *5) * <x >y).

Pattern-matching mode

The ? operator is compatible with any instance at all, like a variable to be assigned; but no variable and no assignment is involved in ?. For example, (>x ? >y) matches (a b c) and (a (b (c)) d), as would be (>x >z >y).

Within a list type instance, a *segment* is a sequence of adjacent elements which are at the same level in the list (the empty sequence is also considered to be a segment: the empty segment). For example, the two element sequence b c is a segment in the instance (a b c); the two element sequence b (c) is a segment in the instance (a (b (c)) d).

When found within a pattern, the operator ?n, where n is an integer, is compatible with any instance segment consisting of precisely n elements. For example, (>x ?2 >y) is compatible with (a b c d). Inside a pattern, the operator * (known as the *absorption operator*) matches any instance segment. (>x * >y), for example, is compatible with both (a b) and (a (c (d e)) f b); in both cases b is assigned to >y. Inside a pattern, the operator *n, where n is an integer, matches any instance segment consisting of at most n elements. For example, (a (>x *5) * <x >y) is compatible with (a ((b c) d e (f g h i)) j (k l) (b c) m); (b c) is assigned to x while m is assigned to y.

Session 2.11
```
(SETQ p1 '(a * b))
= (a * b)

(pattern p1 '(a b))
= t
(pattern p1 '(a c (d e) b f b))
= t
(pattern p1 '(a c (d e) b f c))
= ()
(pattern '(a * b *) '(a c (d e) b f c))
= t
(pattern '(a ? (b *) * c ?) '(a (fie) (b foe foe) c fum))
= t

(SETQ p2 '(surname >x first-names * date-of-birth >z)
     i2 '(surname Smith first-names John Paul date-of-birth 29/2/44))
= (surname smith first-names john henry date-of-birth 29/2/44)

(pattern p2 i2)
= t
x z
= smith
= 29/2/44

(SETQ p3 '(a (b *2 c) ?2))
= (a (b *2 c) ?2)

(pattern p3 '(a (b d c) e f))
= t
(pattern p3 '(a (b d d d c) e f))
= ()
```

Program 2.8

The new parts are underlined.

```
(DE pattern (p i)
   (COND ( (OR (EQ p '?) (EQ p '*)))
          ( (ATOMP p)
            (COND ((EQ p i))
                  ((EQ (caratom p) '>) (SET (cdratom p) i))
                  ((EQ (caratom p) '<) (EQUAL (EVAL (cdratom p))
                   i))))
          ( (EQ (CAR p) '*) (COND ((pattern (CDR p) i))
                                  ((ATOMP i) NIL)
                                  (T (pattern p (CDR i)))))
          ( (AND (ATOMP (CAR p)) (EQ (caratom (CAR p)) '*))
            (COND ((pattern (CDR p) i))
                  ((ZEROP (cdratom (CAR p))) NIL)
                  ((ATOMP i) NIL)
                  (T (pattern (CONS (IMPLODECH (LIST '* (1- (cdratom
                                                            (CAR p)))))
                                    (CDR p))
                             (CDR i)))))
          ( (AND (ATOMP (CAR p)) (NEQ (CAR p) '?) (EQ (caratom (CAR p))
            '?))
            (COND ((< (LENGTH i) (cdratom (CAR p))) NIL)
                  (T (pattern (CDR p) (NTHCDR (cdratom (CAR p)) i)))))
          ((ATOMP i) NIL)
          (T (AND (pattern (CAR p) (CAR i)) (pattern (CDR p) (CDR i))))))

(DE caratom (atom) ; returns the prefix of the atom argument
   (CAR (EXPLODECH atom)))

(DE cdratom (atom) ; returns the radical of the atom argument
   (IMPLODECH (CDR (EXPLODECH atom))))
```

2.4.6   A pattern-matching system third stage
The system described in §§2.4.4 and 2.4.5 is now completed.

Language for patterns

Patterns are now constants, variables, or lists of constants, variables pattern-matching operators and *pattern-matching function calls*. A pattern-matching function call is a list whose first element is a pattern-matching function. There are seven defined *pattern-matching functions*: !and, !or, !not, !liminstan, !evalpat, !outsidepat and !rep. Examples of these 7 functions in use are shown in session 2.12; the reader may wish to look at these while reading the comments below (the session gives at least one illustration of every function).

The format of a call to !and is: (!and {<pattern>}*). This means that !and expects to be followed by a sequence of any number of patterns. Similarly, the format of an !or call is: (!or {<pattern>}*). The format of a

!not call is: (!not <pattern>). The format of a call to !rep (rep for repetition) is: (!rep n <pattern>) where n is an integer.

A call of !liminstan (limitation of instance) can either take the form (!liminstan <variable-to-be-assigned> <procedure>) or the form (!liminstan ? <procedure>). <procedure> is either the name of an already defined LISP procedure, or a LISP lambda-expression (which may use radicals of pattern-matching variables).

The format for a call of !evalpat (pattern evaluation) is (!evalpat <expression>). Similarly, the format of !outsidepat (outside pattern-matching) is (!outsidepat <expression>). <expression> is a LISP expression to be evaluated (which may use radicals of pattern-matching variables).

Pattern-matching mode

If pattern P is of the form (!and {<pattern<}*), P will only be compatible with instance I if it turns out that, one after the other, every one of the <pattern>s following !and is compatible with I. If P is of the form (!or {<pattern>}*), P will be taken to match I as soon as at least one of the <pattern>s following !or is compatible with I. If P is of the form (!not <pattern>), P will only be compatible with I if the <pattern> is not compatible with I.

If P is of the form ((!rep n <pattern>) <rest-of-pattern-P>), P matches I if the pattern (<pattern> ... <pattern> <rest-of-pattern-P>), where <pattern> is repeated n times, is compatible with I.

If P is of the form (!liminstan <variable-to-be-assigned> <procedure>) or (!liminstan ? <procedure>), P is only compatible with I if applying the procedure <procedure> to I returns T (the LISP value T). If the match succeeds and if P is of the form (!liminstan <variable-to-be-assigned><procedure>), I is assigned to <variable-to-be-assigned>. The pattern-matching function !liminstan allows a restriction to be placed on the elements in instances which can be "skipped" by ? or assigned to a variable.

If P is of the form (!evalpat <expression>), P is compatible with I if the new pattern obtained by the LISP evaluation of <expression> matches I. The pattern-matching function !evalpat permits the pattern that will actually be compared with I to be calculated freely at the moment when the pattern-matching takes place. This capacity is especially useful if the final pattern has to depend on what happened earlier on in the pattern-matching (see session 2.12).

If P is of the form ((!outsidepat <expression>) <rest-of-pattern-P>), P is compatible with I if the pattern <rest-of-pattern-P> is compatible with I. In all cases, the LISP evaluation of <expression> is carried out before testing for this compatibility. The pattern-matching function !outsidepat allows a sort of suspension of the pattern-matching in progress while some expression is evaluated. The evaluation of <expression> can make use of already assigned values or it can cause assignments to take place by using radicals of pattern-matching variables (see session 2.12).

Session 2.12

```
(SETQ p4 '(a (!or (b >x) (c >x)) ((!or d e))))
= (a (!or (b x) (c x)) ((!or d e)))

(pattern p4 '(a (c 10) (d)))
= t
x
= 10

(pattern p4 '(a (b 20) (d)))
= t
x
= 20

(SETQ p5 '(>x (!not (a (b <x))) (>y)))
= (>x (!not (a (b <x))) (>y))

(pattern p5 '(fie (a (b fie)) (100)))
= ()

(pattern p5 '(fie (a (b notfie)) (100)))
= t
y
= 100

(SETQ p6 '(a * (!rep 5 b) *))
= (a * (!rep 5 b) *)

(pattern p6 '(a x y b b b b b z))
= t

(pattern p6 'a x y b b b b z u))
= ()

(pattern '(a ((!rep 3 (b * >x)))) '(a ((b c d 7) (b 8) (b e 9))))
= t
x
= 9

(pattern '(>x (!rep 3 <x) *) '(a a a a a a b b))
= t
x
= a

(pattern '(a >x >y (!evalpat (+ <x <y))) '(a 100 200 300))
= t
x y
= 100
= 200

(pattern '( >year
          (!evalpat (COND ((< <year 1968) 'degree) (T 'masters)))
           >subject )
       '(1970 masters science) )
= t
subject
```

```
= science
(DE useful-func (x) (COND ((< x 1968) 'degree) (T 'masters)))
= useful-func

(pattern '(year (!evalpat (useful-func <year>) >subject)
        '(1960 degree arts))
= t
subject
= arts

(DE message (u v) (COND ((< 120 (+ u v)) (PRINT "*** threshold crossed"))))
= message

(pattern '(>x >y (!outsidepat (message <x <y)) ?2 >u) '(50 100 a b c))
*** threshold crossed
= t
u
= c

(SETQ p7
      '( >x
         >y
         (!and (!liminstan ? (LAMBDA (u) (> u 1956)))
               (!liminstan >year (LAMBDA (u) (< u 1966))))
         * ))
= (>x >y (!and (!liminstan ? (lambda (u) (> u 1956))) (!liminstan >year
(lambda (u) (< u 1966)))) * )

(pattern p7 '(Smith John 1962 Exeter 82))
= t
year
= 1962

(pattern p7 '(Jones Peter 1970 Birmingham 31))
= ()

(SETQ p8 '(>x >y (!outsidepat (PRINT "x + y = " (+ x y))) ?2 >z))
= (>x >y (!outsidepat (print "x + y = " (+ x y))) ?2 >z)

(pattern p8 '(50 80 a b c))
x + y = 130
= t
z
= c

(SETQ p9 '(>x (!outsidepat (PROG () (PRINT "key for " x " ?") (SETQ y (READ))))
          * <y >z))
= (>x (!outsidepat (prog () (print "key for " x " ?") (setq y (read)))) * <y >z)

(pattern p9 '(part5 shape2 weight6 quality1 555 $1000))
key for part5 ?
555
= t
z
= $1000
```

(SETQ p10 '(>x (!outsidepat (SETQ x (CAR (EXPLODECH x)))) >y))
= (>x (!outsidepat (setq x (car (explodech x)))) >y)

(pattern p10 '(p45 100))
= t
x y
= p
= 100

Program 2.9

Code and procedures which have been added to program 2.8 are underlined.

```
(DE pattern (p i)
    (COND ( (OR (EQ p '?) (EQ p '*)))
          ( (ATOMP p)
            (COND ((EQ p i))
                  ((EQ (caratom p) '>) (SET (cdratom p) i))
                  ((EQ (caratom p) '<) (EQUAL (EVAL (cdratom p)) i))))
          ( (EQ (CAR p) '*) (COND ((pattern (CDR p) i))
                                  ((ATOMP i) NIL)
                                  (T (pattern p (CDR i)))))
          ( (AND (ATOMP (CAR p)) (EQ (caratom (CAR p)) '*))
            (COND ((pattern (CDR p) i))
                  ((ZEROP (cdratom (CAR p))) NIL)
                  ((ATOMP i) NIL)
                  (T (pattern (CONS (IMPLODECH (LIST '* (1- (cdratom
                                    (CAR p)))))
                                    (CDR p))
                              (CDR i)))))
          ( (AND (ATOMP (CAR p)) (NEQ (CAR p) '?) (EQ (caratom (CAR p))
            '?))
            (COND ((< (LENGTH i) (cdratom (CAR p))) NIL)
                  (T (pattern (CDR p) (NTHCDR (cdratom (CAR p)) i)))))
          (COND ( (EQ (CAR p) '!or) (!or (CDR m)))
                ( (EQ (CAR p) '!and) (!and (CDR m)))
                ( (EQ (CAR p) '!not) (NOT (pattern (CADR p) i)))
                ( (EQ (CAR p) '!liminstan)
                  (COND ((EQ (CADR p) '?) (APPLY (CADDR p) (LIST i)))
                        ((APPLY (CADDR p) (LIST i)) (SET (cdratom (CADR
                         p))i))))
                ( (EQ (CAR p) '!evalpat) (pattern (EVAL (CADR p)) i))
                ( (AND (NOT (ATOMP (CAR p))) (EQ (CAAR p) '!outsidepat))
                  (EVAL (CADAR p))
                  (pattern (CDR p) i))
                ( (AND (NOT (ATOMP (CAR p))) (EQ (CAAR p) '!rep))
                  (pattern (APPEND (!rep (CADAR p) (CADDAR p) NIL) (CDR p)) i))
                ((ATOMP i) NIL)
                (T (AND (pattern (CAR p) (CAR i)) (pattern (CDR p) (CDR i)))))))
```

```
(DE !or (patterns)
    (COND ((NULL patterns) NIL)
          (T (OR (pattern (CAR patterns) i) (!or (CDR patterns))))))

(DE !and (patterns)
    (COND ((NULL patterns))
          (T (AND (pattern (CAR patterns) i) (!and (CDR patterns))))))

(DE !rep (n pattern result)
    (COND ((ZEROP n) result)
          (T (!rep (1− n) pattern (CONS pattern result)))))

(DE caratom (atom)
    (CAR (EXPLODECH atom)))

(DE cdratom (atom)
    (IMPLODECH (CDR (EXPLODECH atom))))
```

We would remind the reader that the pattern-matching system presented here is derived from that of ARGOS-II (see [FARRENY 1980] [PICARDAT 1985]). Other pattern-matching primitives exist in ARGOS-II, in particular variables to which segments can be assigned (of the form: *x) and variables from which segments can be read (of the form: +x).

3

Automatic theorem proving

The representation of knowledge within the framework of formal logics and the mechanisation of the inference modes proposed and analysed by these formal logics constitute one of the original branches of AI. A general paradigm for problem solving consists of (i) encode the individual description as a set of axioms and a conjecture, then (ii) apply a general algorithm to try to derive the conjecture as a theorem based on the axioms and inference modes available; the term *automatic theorem proving* is used to refer to the study and application of this paradigm. After a brief review of the domain, this chapter describes the principles and then the programming of fundamental methods for automatic theorem proving using *first order predicate logic*, based on the resolution principle. Introductory material on the concepts and techniques of automatic theorem proving is to be found in: [NILSSON 1980], [DELAHAYE 1986], [LAURIERE 1986], [FARRENY & GHALLAB 1986]. For more detailed investigation, we recommend in particular: [CHANG & LEE 1973], [LOVELAND 1978], [SIEKMANN & WRIGHTSON 1983] and [WOS 1984].

3.1 REFUTATION BY RESOLUTION STRATEGIES

The reader is presumed to be acquainted with the following notions: *propositions*, *predicates*, *terms*, *atoms*, *literals*, *well formed formulae* (often abbreviated to *formulae* hereafter), *interpretations* and *value of a formula according to an interpretation*. The symbols for *variables*, propositions and predicates are written in upper case; the symbols for *constants* and functions are written in lower case. The *connectives* are denoted by: □ (called the *negation* connective, or *not*), ∨ (*disjunction* or *or*), ∧ (*conjunction* or *and*) and the quantifiers: ∀ (*universal quantifier* or *for all*) and ∃ (*existential quantifier* or *there exists*).

It is assumed that the reader knows how to transform any formula into *clause form*, which means into a *clause conjunction*, which in turn means into a *conjunction of literal disjunctions*. For example, a clause form of the formula:

$$(\forall X) \ (\forall Y) \ \square(P(a,X) \wedge Q(f(Y),b,X)) \wedge R(Y) \quad \text{is:}$$
$$(\square P(a,X) \vee \square Q(f(g(X)),b,X)) \wedge R(g(X))$$

where g is introduced as the symbol for a *Skolem function*. Clause conjunctions are frequently written as *clause sets*. For the previous example this would be:

$$\{\Box P(a,X) \vee \Box Q(f(g(X)),b,X) , R(g(X))\}$$

The reader is also assumed to be acquainted with the notions of *substitutions, compositions of substitutions, unifiers of 2 or more expressions* (an expression is a term or a formula), and *most general unifiers of 2 or more expressions* (abbreviated to **mgu** of 2 or more expressions). Every *elementary* substitution is written in the form: (V t) where t represents the term to be substituted for the variable V, for instance: (Y f(X,c)); non-elementary substitutions are represented by lists of elementary substitutions. The form E.(V t) is used for the result of applying the substitution (V t) to the expression E, for example:

$$\{ M(X,Y) , N(b,U,h(Y)) \}.(Y f(X,c))$$

Set notation is understood in the strict sense: after applying a substitution to a set of expressions, expressions which are now identical are fused into a single element of the new set.

3.1.1   A unification procedure
When applied to a set E of expressions, algorithm 3.1 below will definitely terminate by either finding a mgu of the expressions in E or by discovering that these expressions cannot be unified.

The *disagreement set* of a set E of expressions is obtained in the following manner:
— all the elements of E are scanned left to right, simultaneously, character by character, until the first position where the expressions differ is found (called the *disagreeing* position),
— the expression (term or formula) starting at the disagreeing position is extracted from each element of E,
— the set of these expressions constitutes the disagreement set.

Algorithm 3.1

```
1       PROCEDURE unify(setofexpressions)
2            mgulist <-- empty list
3            UNTIL setofexpressions has just one element REPEAT
4                disagreement <-- disagreement set of setofexpressions
5                IF 2 disagreeing elements V and t do not exist such that
                     V is a variable
                     and t is a term
                     and V does not appear in t
                 THEN RETURN "failure"
6                mgulist <-- mgulist plus the element (V t) at the tail
7                setofexpressions <-- setofexpressions.(V t)
8            ENDUNTIL
9            RETURN mgulist
```

The test 'does V appear in t' (instruction 5) is known as the *occurcheck*. In order to reduce execution costs, certain implementations omit it, thereby incurring the risk of looping. Program 3.1 below is in strict accordance with algorithm 3.1, but it is limited to unifying 2 expressions.

Program 3.1

Terms and atoms are denoted by lists: the function or predicate symbol will be the first element in a list, while the arguments of the function or predicate will be the following elements. A convention for distinguishing variables from constants must be chosen. Here variables will be represented by lists whose first element is the $ symbol (later on the program will be refined to work with variables represented by LISP atoms with $ as the first character: see § 3.2.4.3).

```
(DE unification (exp1 exp2)
    (PROG (disagreement d1 d2 mgulist)
        (COND ((NOT (EQUAL (len exp1) (len exp2))) (RETURN 'failure)))
        (UNTIL (NULL (SETQ disagreement (disag exp1 exp2)))
            (SETQ d1 (CAR disagreement) d2 (CADR disagreement))
            (COND ( (variable? d1)
                    (COND ((in? d1 d2) (RETURN 'failure))
                          ((SETQ mgulist (CONS disagreement mgulist)
                                 exp1 (SUBST d2 d1 exp1)
                                 exp2 (SUBST d2 d1 exp2)))))
                  ( (variable? d2)
                    (COND ((in? d2 d1) (RETURN 'failure))
                          ((SETQ mgulist (CONS
                            (REVERSE disagreement) mgulist)
                                 exp1 (SUBST d1 d2 exp1)
                                 exp2 (SUBST d1 d2 exp2)))))
                  (T (RETURN 'failure))))
        (REVERSE mgulist)))

(DE len (x) ;Called by unification
    (COND ((ATOM x) 1)
          ((EQUAL (CAR x) '$) 1)
          ((LENGTH x))))

(DE disag (exp1 exp2) ;Called by unification
    (COND ( (EQUAL exp1 exp2) NIL)
          ( (OR (ATOM exp1) (ATOM exp2) (EQ (CAR exp1) '$)
            (EQ(CAR exp2) '$))
            (LIST exp1 exp2))
          ( (NOT (EQUAL (CAR exp1) (CAR exp2))) (disag (CAR exp1)
            (CAR exp2)))
          ( ( (disag (CDR exp1) (CDR exp2)))))

(DE variable? (I) ;Called by unification
    (COND ((ATOM I) ()) ((EQUAL (CAR I) '$))))
```

```
(DE in? (l1 l2) ;Called by unification
   (COND ((ATOM l2) NIL)
         ((EQUAL l1 l2))
         ((OR (in? l1 (CAR l2)) (in? l1 (CDR l2))))))
```

Here are some examples of execution:

Session 3.1
(unification '(P (f a) (g ($ y))) '(P ($ x) (g (h b))))
= ((($ x) (f a)) (($ y) (h b)))

(unification '(g ($ x) (f ($ y)) b) '(g a (f (f ($ z))) ($ z)))
= ((($ x) a) (($ y) (f ($ z))) (($ z) b))

(unification '(P (f ($ u) a) (g ($ z) ($ u)) ($ z))
             '(P (f ($ x) ($ y)) (g (h ($ x)) b) (h b)))
= ((($ u) ($ x)) (($ y) a) (($ z) (h ($ x))) (($ x) b))

3.1.2 Paradigm of proof by refutation

Take a finite set of formulae F1 ... Fn representing the various hypotheses extracted from the statement of a problem: the conjunction F1 $\wedge$ ... $\wedge$ Fn takes the truth value **TRUE** for the *interpretation* which corresponds to the problem statement. Let C be the conjecture we are attempting to prove starting from the hypotheses of the statement. It can be shown that C is a *logical consequence* of F1 $\wedge$ ... $\wedge$ Fn (C takes the value **TRUE** for any interpretation which gives the value **TRUE** to F1 $\wedge$ ... $\wedge$ Fn), meaning that $\square$F1 $\vee$ ... $\vee$ $\square$Fn $\wedge$ C is a *valid* formula (takes the value **TRUE** for any interpretation). It is equivalent to show that F = F1 $\wedge$ ... $\wedge$ Fn $\wedge$ $\square$C is an *inconsistent* formula (taking the value **FALSE** for any interpretation); in this case, we are said to proceed *by refutation*.

Let P be a formula and P' a clause form of P (recall that a clause form is a logical conjunction of clauses, or a *set* of clauses). We know that if P is inconsistent then P' is inconsistent, and vice versa. Proving that C is a logical consequence of F1 $\wedge$ ... $\wedge$ Fn is thus equivalent to proving that a clause form F = F1 $\wedge$ ... $\wedge$ Fn $\wedge$ $\square$C is inconsistent. We know too that if a clause form developed for a formula of the type G = G1 $\wedge$ ... $\wedge$ Gm is inconsistent, the conjunction of clause forms developed for each one of the formulae G1 ... Gm respectively is inconsistent, and vice versa; now, by combining the separately developed clause forms for each Gi, we have some chance of reducing the number of Skolem functions introduced when the existential quantifiers are eliminated. Hence, in practice, trying to prove by refutation that C is a logical consequence of F1 $\wedge$ ... $\wedge$ Fn often consists of trying to prove the inconsistency of the conjunction of clause forms F'1, ... F'n, C' obtained separately for F1, ..., for Fn and for $\square$C; the clause form conjunction is a conjunction of clauses that is also frequently written as a *clause set*; this is why trying to prove by refutation that C is a logical consequence of F1 $\wedge$ ... $\wedge$ Fn is treated as attempting to prove the inconsistency of the *set* {F'1, $\wedge$ ..., F'n, C'}.

3.1.3   The resolution principle

The resolution principle is an *inference rule*, meaning a procedure whereby a new formula can be obtained from already existing formulae. It is only defined starting from clauses, these being disjunctions of literals in which all variables are universally quantified (all occurrences of the same variable symbol are governed by the same implicit universal quantifier). For the resolution principle, 2 initial formulae are required: they are called *parent clauses*; a clause derived from applying the resolution principle to 2 clauses, G and H, is called a *resolvent* of G and H.

Let G and H be 2 clauses taken from a clause set (a conjunction of clauses; in practice, the resolution principle is only ever applied to clauses taken from the same set, i.e. taken from a clause conjunction). If the same variable x is present in G and H, all occurrences of x in G (or alternatively in H) can be replaced by any variable name which does not appear in G and H; this is a result of the general logical equivalence of:

$$(\forall X)\ (G(X) \wedge H(X)) \quad \text{and} \quad ((\forall X)\ G(X) \wedge (\forall Y)\ G(Y))$$

This is known as *renaming* variable x. The first thing to be done before attempting to apply the resolution principle to any 2 clauses of a clause set E is to rename as many variables as necessary so that the variable symbols in the various clauses are all distinct.

Let $\{Gi\}_{i=1,p}$ and $\{Hj\}_{j=1,q}$ stand for the literal sets of G and H respectively. Suppose that a subset $\{Gi'\}_{i'=1,p'}$ of $\{Gi\}$ and a subset $\{Hj'\}_{j'=1,q'}$ of $\{Hj\}$ exist such that the set of literals: $\{Gi'\}_{i'=1,p'} \cup \{Hj'\}_{j'=1,q'}$ is unifiable by a mgu r. A resolvent of G and H is obtained as the literal set (clause):

$$(\{Gi\}_{i=1,p}-\{Gi'\}_{i'=1,p'}).r \cup (\{Hj\}_{j=1,q}-\{\square Hj'\}_{j'=1,q'}).r$$

So from:

$$G = P(X,f(a)) \vee P(X,f(Y)) \vee Q(Y) \quad \text{and} \quad H = \square P(Z,f(a)) \vee \square Q(Z)$$

4 distinct resolvents are obtained, including $Q(a) \vee \square Q(X)$.

As another example, $G = \square P(X)$ and $H = P(f(Y)) \vee P(Z)$ give an empty literal set called the *empty clause*. Each application of the resolution principle to 2 particular clauses is termed a *resolution*.

The resolution principle is a *sound* inference rule in the sense that any resolvent is a logical consequence of the 2 parent clauses. As a consequence, if a series of resolutions based on clauses of a set E and resulting resolvents allows the empty clause to be derived, the set E is inconsistent.

The resolution principle is a *refutation complete* inference rule in the sense that if a finite clause set E is inconsistent then there exists a finite sequence of resolutions, based on these clauses and the resolvents derived from them, which leads to the empty clause. Such a sequence defines a *refutation by resolution* of E (often called simply a *refutation*).

The two sets of literals $\{Gi'\}_{i'=1,p'}$ and $\{Hj'\}_{j'=1,q'}$ are said to be *complementary*. If p' = q' = 1, this is called a *binary resolution*; the inference rule defined as a resolution principle restricted to binary resolutions is termed *binary resolution principle*. This rule is not refutation complete.

3.1.4   General algorithm for refutation by resolution

Let us assume we wish to prove that the formula C is a logical consequence of the set of formulae E = {Fi}. Algorithm 3.2 gives the broad outline governing the search for a refutation.

Algorithm 3.2

```
1        proof-by-resolution (E, C)
2            C' <−− set of clause forms for the Fi and for C
3            UNTIL the empty clause belongs to C' REPEAT
4                choose 2 clauses of C', call them C1 and C2
5                IF C1 and C2 have resolvents THEN pick one of them and add
                 it to C
6            ENDUNTIL
```

This algorithm is *nondeterministic* in the sense that it does not specify how C1 and C2 are chosen, nor how a particular resolvent among those of C1, C2 is selected.

This algorithm may not terminate, even if it is true that C is a logical consequence of E. If it does stop the conjecture is proved (since the resolution principle is sound).

The resolution by refutation algorithms examined below are particular (deterministic) examples of algorithm 3.2.

3.1.5   Refutation complete resolution strategies

The way in which the choices anticipated in instructions 4 and 5 of algorithm 3.2 are carried out defines various *resolution by refutation* (or refutation by resolution!) *strategies* (for applying the principle). Certain strategies *restrict* the set of possible resolutions, while others simply *order* them without necessarily prohibiting any of them (though if a strategy produces an infinite series of non-empty clause resolvents when the latter is a logical consequence of the initial clauses, then in actual fact there is a restriction).

A resolution *strategy* is *refutation complete* if, for any finite set E of initial clauses, whenever a refutation by resolution of E exists **there also exists** a refutation by resolution which respects the constraints characterising the strategy. Note that there are often several series of resolutions that respect a given strategy; a strategy can be refutation complete without there being any guarantee that a particular implementation of the strategy is **certain** to find a refutation obeying the strategy, every time such a refutation exists; should such a guarantee exist the strategy will be said to be *directly complete*.

To conclude this section, remember that by using the resolution principle as the only inference rule and by proceeding via refutation, different *theorem provers* can be built, depending on the particular strategy chosen (complete or not). It is equally possible to proceed not by refutation but by *deduction*. A broader range of inference rules can be employed than the resolution principle alone.

3.2 PROLOG-TYPE AUTOMATIC PROVING

3.2.1 General remarks about PROLOG

PROLOG is termed a *declarative* programming language to highlight the contrast with classical *imperative* (or *procedural*) languages. In effect, the PROLOG user is limited to a certain degree (greater than in classical languages) to declaring those application-dependent relationships which interest him; he uses logical clauses to encode the hypotheses in the statement of the problem and the conjecture being studied; thereafter it is the language *interpreter* which undertakes (or attempts) to prove that the conjecture is a logical consequence of the hypotheses using refutation and resolution. In practice, however, PROLOG dialects also offer imperative-type primitives for tasks such as assignments or input–output.

An interpreter of a perfectly declarative language ought to demonstrate the same problem-solving powers regardless of how the declarations of the hypotheses and conjecture were laid out. This is not so in PROLOG (nor clearly for any classical programming language). The PROLOG interpreter takes the hypotheses and the conjecture to be coded as a *sequence* (and not an unordered set) of Horn clauses; these are clauses which can include at most one *positive literal* (a literal is positive is it does not include the negation connective). The strategy used by the interpreter is sensitive to this ordering of initial clauses: the order of the clauses characterising a problem can determine whether or not the interpreter will solve it.

The refutation by resolution strategy implemented by PROLOG is a particular (more constrained) adaptation of a type of strategy known as *LUSH resolution* (LUSH stands for: Linear resolution with Unrestricted Selection function for Horn clauses). LUSH type strategies are special cases of *linear input strategies*, which are themselves special cases of *linear strategies*. The linear strategies are refutation complete (by normal resolution); linear input strategies are not complete in general terms but are refutation complete for binary resolution if the sets of initial clauses are restricted to Horn clauses; LUSH strategies, being defined only for sets of Horn clauses, are also complete for refutation by **binary** resolution. Further details about these families of strategies and their interrelationships are to be found in the references cited at the beginning of the chapter; § 3.2 only describes PROLOG's strategy (which is incomplete).

Finally we should remember that although the theoretical foundations of PROLOG are based on first order predicate logic restricted to Horn clauses, in practice (and at his own risk!) the PROLOG programmer can manipulate *non-Horn* clauses or even get into higher order logics, such as using variable predicates. With regard to the art of PROLOG programming, see: [CLOCKSIN & MELLISH 1985], [DONZ & HURTADO 1985], [GIAN-NESINI *et al.* 1985], [CONDILLAC 1986] and, in part, [FARRENY & GHALLAB 1986].

3.2.2   PROLOG-type refutation by resolution strategy

A PROLOG program is a sequence of Horn clauses written in accordance with a syntax which varies between dialects. The program proper is a series of clauses, P, each one of which has exactly one positive literal; this guarantees that the clause set is consistent (an interpretation satisfying the set does exist). To run the program the user proposes a clause, C, without any positive literal; such a clause represents a formula of the type: (∀X) ... (∀Z) □P(...) ∨ ... □R(...) or alternatively the **negation** of a formula, C', of the type: (∃X) ... (∃Z) P(...) ∧ ... ∧ R(...). The interpreter then tries to prove that the union of P's clauses with C forms an inconsistent set, which means that C' is a logical consequence of P. Clause C is frequently called a *query* or *list of goals*.

As an example, let us use clauses and the syntax introduced at the start of the chapter to represent the following problem: given the facts that Mark is fair-haired, John is dark-haired, Peter is John's father, Mark is Peter's father, John is Robert's father, that someone is someone else's father if the latter is the former's son, that someone is someone else's grandfather (call this someone else p) if the former is someone's father and this someone is p's father, and finally that Mark is George's son, we wish to prove that there is some fair-haired grandfather.

Set of clauses representing the hypotheses and the negation of the conjecture:

> { FAIR(mark), DARK(john), FATHER(peter,john),
>   FATHER(mark,peter),
>   FATHER(john,robert), FATHER(X,Y) ∨ □SON(Y,X),
>   GRANDFATHER(X,Y) ∨ □FATHER(X,Z) ∨ □FATHER(Z,Y),
>   SON(mark,george),
>   □GRANDFATHER(X,Y) ∨ □FAIR(X) }

Here is how this problem is coded in *PROLOG-II*, the dialect of PROLOG originally developed at the University of Marseille and available on many computers:

Program proper:

```
fair(mark);
dark(john);
father(peter,john);
father(mark,peter);
father(john,robert);
father(x,y) -> son(y,x);
grandfather(x,y) -> father(x,z) father(z,y);
son(mark,george);
```

Request (or list of goals):

```
grandfather(u,v) fair(u);
```

Note: the query representing the clause:

□GRANDFATHER(X,Y) ∨□FAIR(X),

where variables X and Y are universally quantified, can also be read as: ∃? u and v such that grandfather(u,v) ∧ fair(u).

We now present the first version of the code which we intend to use for this problem so that it can be submitted to the first PROLOG interpreter which we shall write (program 3.2); variables are lists beginning with $; the second PROLOG interpreter (program 3.3) will accept a simpler form:

Program proper:

```
( ((fair mark))
  ((dark john))
  ((father peter john))
  ((father mark peter))
  ((father john robert))
  ((father ($ x) ($ y)) (son ($ y) ($ x)))
  ((grandfather ($ x) ($ y)) (father ($ x) ($ z)) (father ($ z) ($ y)))
  ((son mark george)) )
```

Request (or list of goals):

```
( (grandfather ($ u) ($ v)) (fair ($ u)) )
```

This form of coding will be used until otherwise indicated.

Let us review the principle of how a PROLOG interpreter functions. The list of goals is kept as a stack. The interpreter cycles round trying to apply the resolution principle between the query clause and one of the program clauses, examining them in the order they are found in the program; to be more precise, it attempts to unify the top of the stack, which is the first (negative) literal in the query clause, with the positive literal of a program clause; in the preceding example (see the above code), the first cycle will unify the first goal with the positive literal of the eighth element (i.e. the eighth clause) in the program-list (notice that the positive literals appear at the head of each clause).

Assume that such a resolution is possible. The resolvent obtained will act as the list of goals for the next cycle: the choice of resolvents to be compared with the program clauses (which is fixed in basic PROLOG) is driven in a depth first fashion. In accordance with the definition of the resolution principle, the mgu obtained is applied at the time of the unification of the literals complementary to the other literals implied in the resolution, and these are joined together to form the resolvent; more precisely, the literals arising from the resolved program clause are put in front of the remaining goals, maintaining their mutual order. At the end of the first cycle in the above example, the resolvent has the following form (with the exception of variable names which were renamed before any resolution was performed):

```
((father ($ x) ($ z)) (father ($ z) ($ y)) (fair ($ x)))
```

If it is not possible to resolve the query clause (via its first literal) in combination with a program clause, the interpreter carries out a backtrack: it restores the goal list which existed at the beginning of the preceding cycle and tries another resolution with those program clauses which follow the one previously reached, still following the order of the program.

The way a PROLOG interpreter exploits the declarative knowledge stored in each PROLOG program naturally evokes the principles of expert systems. A PROLOG program combines the base of established facts — clauses reduced to one (positive) literal and the rule base, which is made up of the other clauses — into a unified knowledge base. The most important original feature of the knowledge expression language, in comparison with that used by the majority of expert systems, is that it permits variables in facts as well as in rules. The list of goals represents a base of facts to be established (or, equally, to be refuted). Each rule (corresponding to a Horn clause) states that the fact represented by the first literal is established when the facts represented by the others literals are established. The rules are thus invoked by the first literal in a backward-chaining fashion. The compatibility test between the current head of the base of facts to be established and the trigger of each rule consists of an attempt at unification in the sense of section 3.1.1. In summary, PROLOG interpreters are inference engines which work using backward- chaining and depth first strategies, are able to backtrack, and allow variables in both facts and rules.

Algorithm 3.3 formalises the mode of operation summarised above. The main procedure, justify, is presumed to be called initially by: justify(query, program) once the lists query and program have been loaded. The unify procedure is the one in algorithm 3.1 (§ 3.1.1), except that here it requires two arguments: the two expressions to be unified.

Algorithm 3.3

1	PROCEDURE justify(goalstack, restprog)
2	VARIABLES agoal, anitemofknowledge, theconclusion substitutions1, substitutions2, goals
3	IF goalstack is empty THEN RETURN an empty list
4	IF restprog is empty THEN RETURN "failure"
5	agoal <−− top of goalstack
6	anitemofknowledge <−− first element of restprog, with renamed variables
7	theconclusion <−− positive literal of anitemofknowledge
8	substitutions1 <−− unify(agoal, theconclusion)
9	IF substitutions1 is not equal to "failure" THEN
9.1	goals <−− result of successively applying the substitutions in the list substitutions1, in the order given, to the result of concatenating goalstack, except for agoal, to the tail of the list of negative literals of anitemofknowledge
9.2	substitutions2 <−− justify(goals, program)
9.3	IF substitutions2 is not equal to "failure" THEN RETURN the list of substitutions obtained by concatenating the list substitutions2 to the tail of the list substitutions1
9.4	ENDIF
10	restprog <−− restprog minus anitemofknowledge
11	RETURN justify(goalstack, restprog)

3.2.3   A first PROLOG interpreter

The section of program 3.2 below starting at procedure *justify* follows algorithm 3.3 closely. The *unification* procedure called by *justify* was described in § 3.1.1 (program 3.1), so neither this procedure nor any of those which depend on it is repeated here. The main procedure is microprolog and it serves as a simple interface between the user and the actual demonstrator represented by justify. To make it easier to understand, session 3.3 illustrates that justify may be called directly; session 3.3 shows the interface being used. The display of results is left in its raw state for the time being: this allows the reader to follow the substitutions undergone by all the variables implied in the resolution, whether or not they belong to the query; another procedure for displaying results is suggested later on.

Program 3.2

```
(DE microprolog () ;Main procedure normally called by the user
    (PROG (command)
        (UNTIL NIL
            (prompt "next command")
            (SETQ command (READ))
            (COND ( (EQ 'stop command) (RETURN 'bye))
                ( (EQ '? command)
                  (PRINT "The commands understood are:")
                  (PRINT "newprog oldprog newqry oldqry")
                  (PRINT "saveprog saveqry eval stop and ?"))
                ( (EQ 'newprog command)
                  (prompt "text of program") (SETQ program (READ)))
                ( (EQ 'oldprog command)
                  (prompt "identifier") (SETQ program (EVAL (READ))))
                ( (EQ 'newqry command)
                  (prompt "text of query") (SETQ query (READ)) (invoke))
                ( (EQ 'oldqry command)
                  (prompt "identifier") (SETQ query (EVAL (READ)))
                  (invoke))
                ( (EQ 'saveprog command)
                  (prompt "what name") (SET (READ) program))
                ( (EQ 'saveqry command) (prompt "what name")
                                        (SET (READ) query))
                ( (EQ 'eval command) (prompt "expression")
                                     (PRINT (EVAL (READ))))
                (T (PRINT "unknown command"))))))
(DF prompt message ;Called by microprolog
    (TERPRI) (PRINT (CAR message) "=>"))

(DE invoke () ;Called by microprolog
    (PROG (index result)
        (SETQ index 0 result (justify query program))
        (COND ( (EQ 'failure result)
                (TERPRI) (PRINT "I can't find any justification"))
            ( (NULL result) (TERPRI) (PRINT "that's correct !"))
            (T (TERPRI)
                (PRINT "*** substitutions needed to satisfy your query:")
                (PRINT result)))))
```

```
(DE justify (goalstack restprog)
;Called by invoke and justify, it conforms to algorithm 3.3
    (PROG (substitutions1 substitutions2 anitemofknowledge)
        (COND
            ((NUL goalstack) NIL)
            ((NULL restprog) 'failure)
            ((AND
                (NEQ 'failure
                    (SETQ substitutions1
                        (unification
                          (CAR goalstack)
                          (CAR (SETQ anitemofknowledge
                                    (rename (CAR restprog)))))))
                (NEQ 'failure
                    (SETQ substitutions2
                        (justify
                          (applysubstit
                            substitutions1
                            (APPEND (CDR anitemofknowledge)
                                    (CDR goalstack)))
                          program))))
            (APPEND substitutions1 substitutions2))
          (T (justify goalstack (CDR restprog))))))

(DE rename (expr) ;Called by justify
    (SETQ index (1+ index)) (renam1 expr))

(DE renam1 (expr) ;Called by rename and renam1
    (COND ((ATOM expr) expr)
          ((EQUAL (CAR expr) '$) (LIST '$ (CADR expr) index))
          ((MAPCAR 'renam1 expr))))

(DE applysubstit (listsubs listexpr) ;Called by justify
    (UNTIL (NULL listsubs)
        (SETQ listexpr (SUBST (CADAR listsubs) (CAAR listsubs) listexpr))
        (NEXTL listsubs))
    listexpr)
```

Session 3.2 is obtained after loading program 3.2 (apart from **micro-prolog** and **invoke**).

Session 3.2
```
(SETQ program
    '( ((fair mark))
       ((dark john))
       ((father peter john))
       ((father mark peter))
       ((father john robert))
       ((father ($ x) ($ y)) (son ($ y) ($ x)))
       ((grandfather ($ x) ($ y)) (father ($ x) ($ z)) (father ($ z) ($ y)))
       ((son mark george)))
    index 0)
```

= 0
(justify '((grandfather ($ u) ($ v)) (fair ($ u))) program)
= ((($ u) ($ x 7)) (($ v) ($ y 7)) (($ x 7) mark) (($ z 7) peter) (($ y 7) john))
(justify '((grandfather ($ u) peter)) program)
= ((($ u) ($ x 46)) (($ y 46) peter) (($ x 46) ($ x 100)) (($ z 46) ($ y 100)) (($ y 100) mark) (($ x 100) george))

In order to systematically rename variables in a unique manner, pro-cedures **rename** and **renam1** introduce a new number to the tail of the lists representing variables; internal variables thus have the form: ($ x 1), ($ x 2) etc. ... The variable **index**, which is initialised to 0 here, serves as a reference point for generating new numbers. The value returned by **justify** is the list of all the substitutions performed. After the first call of **justify**, it can be seen that ($ u) has been replaced by ($ x 7) and this in turn replaced by **mark**, while ($ v) has been replaced by ($ y 7) and then this replaced by **john**; in addition, the variable ($ z 7) which was introduced in the course of searching has been replaced by **peter**. By the end, what interests the user are the final substitutions that u and v undergo: (u mark) and (v john), which are the only variables present in the query.

We have chosen to blindly rename all the variables of the clause in the PROLOG program which the interpreter is trying to resolve with the current top of the goal stack. One could instead attempt to rename only those variables in the clause which already figure in the top of the goal stack, but such a procedure means that before every resolution one has to determine what variables are common to the two expressions to be resolved.

Session 3.3 is obtained after loading program 3.2.

Session 3.3
(microprolog)

next command =>
?
The commands understood are:
newprog oldprog newqry oldqry
saveprog saveqry eval stop and ?

next command =>
newprog

text of program =>
(((fair mark))
     ((dark john))
     ((father peter john))
     ((father mark peter))
     ((father john robert))
     ((father ($ x) ($ y)) (son ($ y) ($ x)))
     ((grandfather ($ x) ($ y)) (father ($ x) ($ z)) (father ($ z) ($ y)))
     ((son mark george)))

```
next command =>
newqry
```

```
text of query =>
((grandfather ($ u) ($ v)) (fair ($ u)))
```

*** substitutions needed to satisfy your query:
((($ u) ($ x 7)) (($ v) ($ y 7)) (($ x 7) mark) (($ z 7) peter) (($ y 7) john))

```
next command =>
saveqry
```

```
what name =>
query1
```

```
next command =>
newqry
```

```
text of query =>
((grandfather ($ u) peter))
```

*** substitutions needed to satisfy your query:
((($ u) ($ x 7)) (($ y 7) peter) (($ x 7) ($ x 61)) (($ z 7) ($ y 61))
(($ y 61) mark) (($ x 61) george))

```
next command =>

eval
```

```
expression =>
query1
((grandfather ($ u) ($ v)) (fair ($ u)))
```

```
next command =>
oldqry
```

```
identifier =>
query1
```

*** substitutions needed to satisfy your query:
((($ u) ($ x 7)) (($ v) ($ y 7)) (($ x 7) mark) (($ z 7) peter) (($ y 7) john))

```
next command =>
stop
= bye
```

All queries (lists of goals) invoked by the user via the keywords **newqry** and **oldqry** refer to the last program (*current program*) introduced by **newprog** or **oldprog**. When he wants to construct a PROLOG program, the user first enters the keyword **newprog**; by using **saveprog**, he can save the current program, assigning it an identifier; if the user wants a previously written program to become the current program, he uses the keyword **oldprog** and then supplies that program's identifier. In the same way, the user can construct a new query after **newqry** or supply the identifier of an already written query after **oldqry**; the last query entered can be saved (keyword **saveqry**). The user is also able to ask for the value of any LISP expression without leaving **microprolog** (keyword **eval**). For instance, by asking for **program** or **query**, which are global variables in program 3.2, to be evaluated, the user can gain access to the text of the current program or the text of the last query.

Three improvements to program 3.2 are now suggested, and they will be incorporated into program 3.3 later on.

3.2.4 A second PROLOG interpreter

3.2.4.1 Modification of the input dialogue

The way the keywords known to program 3.2 are written is deliberately very explicit; this can easily be changed, and the interface program (microprolog) can be more generally adapted as one pleases. In the version of microprolog below the keywords which the user must enter have been reduced to mnemonics, and in particular there is no longer any need to announce explicitly that a new query is to be put: any command not preceded by a keyword is presumed to be a query (list of goals). As before, however, no syntactic checking of programs or queries is carried out; furthermore, unknown commands are not now detected as such but rather treated as queries.

```
(DE microprolog () ;Main procedure normally called by the user
    (PROG (command)
        (UNTIL NIL
            (prompt "YOUR COMMAND")
            (SETQ command (READ))
            (COND ( (EQ 'S command) (RETURN 'bye))
                  ( (EQ '? command)
                    (PRINT "The commands understood are:")
                    (PRINT "NP,OP (New or Old Program), OQ (Old
                      Query),")
                    (PRINT "SP,SQ (Save Program or Query),")
                    (PRINT "E (LISP Evaluation), S (Stop) and ?"))
                  ( (EQ 'NP command) (prompt "text of New Program ?")
                    (SETQ program (READ)))
                  ( (EQ 'OP command) (prompt "name of Old Program ?")
                    (SETQ program (EVAL (READ))))
                  ( (EQ 'OQ command) (prompt "name of Old Query ?")
                    (SETQ query (EVAL (READ))) (invoke))
                  ( (EQ 'SP command)
                    (prompt "Save Program by what name ?")
                    (SET (READ) program))
                  ( (EQ 'SQ command) (prompt "Save Query by what
                  ( name ?")
                    (SET (READ) query))
                  ( (EQ 'E command) (prompt "Evaluate what
                    expression ?")
                    (PRINT (EVAL (READ))))
                  (T (SETQ query command) (invoke))))))
```

This new version of the interface will be seen in action from session 3.4 onwards.

3.2.4.2 Modification of the output of results

The procedure invoke can be modified as follows so that only the *final* substitutions of variables *appearing in the queries* are displayed:

```
(DE invoke () ;Called by microprolog
    (PROG (index result)
        (SETQ index 0 result (justify query program))
        (COND (  (EQ 'failure result)
                 (TERPRI) (PRINT "I can't find any justification"))
              (  (NULL result) (TERPRI) (PRINT "that's correct !"))
              (T (TERPRI)
                 (PRINT "*** substitutions needed to satisfy your query:")
                 (format (globalsortout result NIL))))))
```

with:

```
(DE globalsortout (exlist newlist) ;Called by invoke and globalsortout
    (COND (  (NULL exlist) (REVERSE newlist))
          (  (NOT (NULL (CDDAAR exlist)))
             (globalsortout (CDR exlist)
                     (applysubstit (LIST (CAR exlist)) newlist)))
          (T (globalsortout (CDR exlist) (CONS (CAR exlist) newlist)))))
```

and

```
(DE format (listsubs) ;Called by invoke
    (PRIN '{)
    (MAPC (LAMBDA (x) (PRIN '| | (CAR x)) (PRIN '=) (PRIN (CADR x) '| |))
          listsubs)
    (PRINT '}))
```

Here is an example of the work performed by globalsortout:

```
(globalsortout
    '((($ u) ($ x 7)) (($ v) ($ y 7)) (($ x 7) mark) (($ z 7) peter) (($ y 7) john)) NIL)
= ((($ u) mark) (($ v) john))
```

Note: the way globalsortout is written takes advantage of the facts that:

— the lists representing variables present in the query only have 2 elements while those representing variables introduced by the resolution have 3 elements,
— the variables appearing in the query are never replaced by other variables, no matter what they are (this is a result of the way the unification procedure is defined and called: justify always passes the goal to be unified to unification as the *first* argument and unification does not treat both arguments in the same manner); for this reason there is no need, in the last line of globalsortout, to apply applysubstit to result, unlike what is done in the penultimate line.

The role of format is to simplify how results are presented. A result such as: ((($ u) mark) (($ v) john)) will be rewritten as:

$$\{ (\$ u) = mark (\$ v) = john \}.$$

3.2.4.3 Modification of the external representation of variables

From an algorithmic point of view, the distinction between the symbols for variables and the symbols for other objects in the language for writing clauses (constants, functions, predicates, connectives) is made in the body of the unification procedure (unification). However, the convention adopted up until now (a $ character at the head of a *list*, the second element being the variable identifier) is somewhat tedious for the user, both during input and output. It would undoubtedly be preferable to denote variables by *atoms* whose first character might still be, say, $.

The first way to cope with such variables is to look out for a $ at the head of certain atoms *in the unification procedure*. This was what we did in the pattern-matching procedures given at the end of chapter 2. In this situation, atoms have to be exploded into characters (by EXPLODECH) every time unification is attempted, which can prove cumbersome.

Here we have preferred to transform the external atomic form of variables, $x or $expression for instance, into list form, ($ x) or ($ expression) respectively, *at the input stage*. When results are displayed variables will revert to the external form. Hence the number of calls to EXPLODECH will be substantially reduced.

To transform the atomic form of each variable in an expression into list form, explode could be defined as:

```
(DE explode (expression)
    (PROG (list)
        (COND (  (ATOMP expression)
                 (SETQ list (EXPLODECH expression))
               (   (COND ((EQ '$ (CAR list)) (LIST '$ (IMPLODECH (CDR list)))))
               (       (T expression)))
            (T (MAPCAR (LAMBDA (x) (explode x)) expression)))))
```

As an example:

```
(explode '(a $a3 b $value (c $term90)))
= (a ($ a3) b ($ value) (c ($term90)))
```

Rather than using explode, it is in fact simpler to complete program 3.2 by defining $ as a macro-character (a macro-character encountered in the input stream is replaced by the value returned from executing the function associated with the macro-character at the time it is defined):

```
(DMC $ () (LIST '* (READ)))
```

Note that it is arranged to replace $x or $expression by (* x) or (* expression) respectively; this avoids reintroducing $ into the transformed programs or queries so that fresh, unexpected transformations do not take place when previously transformed programs or queries are manipulated later on. A procedure, paste, is introduced to display results:

```
(DE paste (expression) ;Called by invoke and paste
    (COND ((ATOMP expression) expression)
          ((EQ '* (CAR expression)) (APPLY 'CONCAT
          (CONS '|$| (CDR expression))))
          (T (MAPCAR (LAMBDA (x) (paste x)) expression))))
```

and the last line of invoke is changed to:

```
(format (paste (globalsortout result NIL)))))))
```

As an example:

```
(paste '(a (* a 3) b (* value) (c (* term 9 0))))
= (a $a3 b $value (c $term90))
```

Lastly, program 3.3 below is derived from program 3.2 with the integration of the 3 improvements that have just been described (§ 3.2.4.1 to 3.2.4.3); in order to save space, neither the procedure unification nor any of its associated utilities is shown (see program 3.1).

Program 3.3

```
(DE microprolog () ;Main procedure normally called by the user
    (PROG (command)
        (UNTIL NIL
            (prompt "YOUR COMMAND")
            (SETQ command (READ))
            (COND ( (EQ 'S command) (RETURN 'bye))
                  ( (EQ '? command)
                    (PRINT −The commands understood are:−)
                    (PRINT "NP,OP (New or Old Program), OQ (Old Query),")
                    (PRINT "SP,SQ (Save Program or Query)")
                    (PRINT "E (LISP Evaluation), S (Stop) and ?"))
                  ( (EQ 'NP command) (prompt "text of New Program ?")
                    (SETQ program (READ)))
                  ( (EQ 'OP command) (prompt "name of Old Program ?")
                    (SETQ program (EVAL (READ))))
                  ( (EQ 'OQ command) (prompt "name of Old Query ?")
                    (SETQ query (EVAL (READ))) (invoke))
                  ( (EQ 'SP command)
                  ( (prompt "Save Program by what name ?")
                    (SET (READ) program))
                  ( (EQ 'SQ command)
                  ( (prompt "Save Query by what name ?")
                    (SET (READ) query))
                  ( (EQ 'E command)
                  ( (prompt "Evaluate what expression ?")
                    (PRINT (EVAL (READ))))
                  ( (SETQ query command) (invoke)))))))
(DF prompt message ;Called by microprolog
    (TERPRI) (PRINT (CAR message) " => "))
(DE invoke () ;Called by microprolog
    (PROG (index result)
        (SETQ index 0 result (justify query program))
        (COND ( (EQ 'failure result)
                (TERPRI) (PRINT "I can't find any justification"))
              ( (NULL result) (TERPRI) (PRINT "that's correct !"))
              (T (TERPRI)
                 (PRINT "*** substitutions needed to satisfy your query:")
                 (format (paste (globalsortout result NIL)))))))
```

```
(DE globalsortout (exlist newlist) ;Called by invoke and globalsortout
     (COND (  (NULL exlist) (REVERSE newlist))
           (  (NOT (NULL (CDDAAR exlist)))
              globalsortout (CDR exlist)
                        (applysubstit (LIST (CAR exlist)) newlist)))
           (T (globalsortout (CDR exlist) (CONS (CAR exlist) newlist)))))

(DMC $ () (LIST '* (READ)))

(DE paste (expression) ;Called by invoke and paste
     (COND ((ATOMP expression) expression)
           ((EQ '* (CAR expression))
            (APPLY 'CONCAT (CONS '|$| (CDR expression))))
           (T (MAPCAR (LAMBDA (x) (paste x)) expression))))

(DE format (listsubs) ;Called by invoke
     (PRIN '{)
     (MAPC (LAMBDA (x) (PRIN '| | (CAR x)) (PRIN '=) (PRIN (CADR x) '| |))
           listsubs)
     (PRINT '}))

(DE justify (goalstack restprog) ;Called by invoke and justify
     (PROG (substitutions1 substitutions2 anitemofknowledge)
          (COND
               ((NUL goalstack) NIL)
               ((NULL restprog) 'failure)
               ((AND
                    (NEQ 'failure
                         (SETQ substitutions1
                              (unification
                               (CAR goalstack)
                               (CAR (SETQ anitemofknowledge
                                         (rename (CAR restprog)))))))
                    (NEQ 'failure
                         (SETQ substitutions2
                                   (justify
                                   (applysubstit
                                        substitutions1
                                        (APPEND (CDR anitemofknowledge)
                                                   (CDR goalstack)))
                                   program))))
                (APPEND substitutions1 substitutions2))
               (T (justify goalstack (CDR restprog))))))

(DE rename (e) ;Called by justify
     (SETQ index (1+ index)) (renam1 e))

(DE renam1 (e) ;Called by rename and renam1
     (COND ((ATOM e) e)
           ((EQUAL (CAR e) '*) (LIST '* (CADR e) index))
           ((MAPCAR 'renam1 e))))
```

```
(DE applysubstit (listsubs listexpr) ;Called by justify and globalsortout
    (UNTIL (NULL listsubs)
        (SETQ listexpr (SUBST (CADAR listsubs) (CAAR listsubs) listexpr))
        (NEXTL listsubs))
    listexpr)
```

Session 3.4 is obtained **after** the **whole** of program 3.3 has been loaded (hence including the definition of the macro-character $). In this example session it was decided to initialise 3 identifiers (**program1**, **qry1** and **qry2**) *before* invoking **microprolog**; these identifiers could have been initialised from **within** microprolog, as in session 3.3. Be careful: should you wish to recover the value of identifiers set up **before** the macro-character $ was defined and containing variables from within **microprolog**, you may be surprised by the result.

Session 3.4
```
(SETQ
    program1
    '( ((fair mark)) ((dark john)) ((father peter john))
        ((father mark peter)) ((father john robert)) ((father $x $y) (son $y $x))
        ((grandfather $x $y) (father $x $z) (father $z $y)) ((son mark george)))
    qry1
    '((grandfather $u $v) (fair $u))
    qry2
    '((grandfather $u peter)))
= ((grandfather $u peter))

(microprolog)

YOUR COMMAND =>
OP

name of Old Program ? =>
program1

YOUR COMMAND =>
OQ

name of Old Query ? =>
qry1

*** substitutions needed to satisfy your query:
{ $u=mark $v=john }

YOUR COMMAND =>
OQ

name of Old Query ? =>
qry2

*** substitutions needed to satisfy your query:
{ $u=george }
```

```
YOUR COMMAND =>
((father $u $v))
```

*** substitutions needed to satisfy your query:
{ $u=peter $v=john }

```
YOUR COMMAND =>
SQ
```

Save Query by what name ? =>
qry3

```
YOUR COMMAND =>
E
```

Evaluate what expression ? =>
qry3
((father (* u) (* v)))

```
YOUR COMMAND =>
((father $x $y) (dark $x))
```

*** substitutions needed to satisfy your query:
{ $x=john $y=robert }

```
YOUR COMMAND =>
((father george $x))
```

*** substitutions needed to satisfy your query:
{ $x=mark }

```
YOUR COMMAND =>
((grandfather george $x))
```

*** substitutions needed to satisfy your query:
{ $x=peter }

```
YOUR COMMAND =>
S
= bye
```

Note: by calling for the identifier qry3 to be evaluated we get to see its internal form; to get the variables to be rewritten in external form we have simply to ask for (paste qry3) to be evaluated in place of qry3.

Program 3.3 stops as soon as one set of substitutions satisfying the query is found: this is also the case after the query qry3 in the preceding session, although there are several possible responses.

3.2.5 Third PROLOG interpreter: searching for multiple proofs

Unlike the two prototype PROLOG interpreters we have seen so far, proper PROLOG interpreters have the ability to reinvoke the search for proofs after the first one has been found. Many interpreters, for example those for the PROLOG-II dialect (GIANNESINI *et al.* 1985), automatically reinvoke the search as justifications are found: it is up to the user to interrupt this mechanism if he wants to do so, in particular by using the predefined predicate / which will be discussed later on. Other interpreters, such as those for the Edinburgh PROLOG dialect (CLOCKSIN & MELLISH 1985), look for just one proof initially; if the user wants a new justification it is up to him to say so (for example, if the user just presses the carriage-return when given a reply this means that he does not want any more, while pressing the ';' character sets off a search for a new reply).

The interpreter represented by program 3.4 below can work in two ways. If the interpreter is initially called by (microprolog t), it will automatically reinvoke the search for new solutions; if the initial call is (microprolog), it will only try to find a single solution automatically (but after the interpreter has found one solution the user will be able to request a search for a new one).

The principle for transforming a PROLOG interpreter capable of searching for a first proof into an interpreter which can search for new proofs is relatively simple. All that has to be done is that, rather than stopping when the goal stack is empty, the interpreter should instead continue the search, as it already does when restprog (see the text of justify), the set of clauses still to be compared with the top of the goal stack, is empty. Of course, the solution must be displayed at the time it is found, which means printing the list LS of the substitutions which allowed the goal stack to be emptied (or at least that part of LS dealing with the variables present in the query).

This last operation is difficult to perform from within the justify procedure of program 3.3 (or program 3.2) since at the moment when goalstack is found to be empty (line 3 of justify), the list LS is not available. When goalstack becomes empty, the LS list is actually constructed via successive executions of (APPEND substitutions1 substitutions2) which correspond to going back up the sequence of calls to justify. Therefore, *the list LS is only constructed once we have gone back up the sequence of calls to justify* and reached the first call (from invoke).

Below is a new version of justify which makes the list LS available from the instant when the stack of goals is found to be empty. Here, the justify procedure can only return one or other of the two values failure (line 4 and last line) or success (line 3 and last line). It now accepts a third argument, allsubst (all substitutions), which is called with the list of substitutions performed up to this point; when justify is called for the first time, in invoke, allsubst is called with NIL. When a resolution between the top of the goal stack and the head of the first clause in the current program (restprog) succeeds, the list allsubst plus the list resultunif (result of unification) of substitutions necessary for the unification is supplied as the third argument to a call of justify which is applied to the goal stack after resolution. The only way for this call of justify to return success would be via an escape route (named finished and declared in invoke), in which case the return would take place in invoke and not in the calling justify; to put it another way, if after the call to justify appearing in lines 9 to 13 we return into the calling procedure justify, this can only be because the attempt to resolve the new goal stack (the CDR of goalstack) has failed; as anticipated in the final line, it is then necessary to set off a new call to justify starting with the initial goalstack, decrementing the current program (restprog) by one instruction and restarting from the list of substitutions already recorded, allsubst.

Note that it would be easy to do without TAG and EXIT: it would only be necessary to substitute (RETURN 'success) for (EXIT finished 'success), to store the value from the call of justify in lines 9 to 13 in a new local variable, resultjustif, and to insert the following line just before the last line: (COND ((EQ 'success resultjustif) (RETURN 'success))).

```
(DE justify (goalstack restprog allsubst) ;Called by invoke and justify
    (PROG (resultunif anitemofknowledge)
         (COND ( (NULL goalstack) (writesolution allsubst)
                                  (EXIT finished 'success))
               ( (NULL restprog) (RETURN 'failure)))
         (SETQ resultunif
              (unification (CAR goalstack)
                           (CAR (SETQ anitemofknowledge
                                      (rename (CAR restprog))))))
         (COND ( (NEQ 'failure resultunif)
                 (justify
                    (applysubstit
                        resultunif (APPEND (CDR anitemofknowledge)
                                           (CDR goalstack)))
                    program
                    (APPEND allsubst resultunif))))
         (justify goalstack (CDR restprog) allsubst)))

(DE invoke () ;Called by microprolog
    (PROG (index)
         (SETQ index 0)
         (COND ( (EQ 'failure (TAG finish (justify query program NIL)))
                 (TERPRI) (PRINT "I can't find any justification")))))

(DE writesolution (allsubst) ;Called by justify
    (COND ((NULL allsubst) (TERPRI) (PRINT "that's correct !"))
          (T (TERPRI)
             (PRINT "*** substitutions needed to satisfy your query:")
             (format (paste (globalsortout allsubst NIL))))))
```

invoke is now divided into two parts; the part which has inherited the name invoke makes the initial call to justify and issues a message in the case of total failure. When the expression beginning with TAG is evaluated, if the escape called finished is taken this can only come from evaluating (EXIT finished 'success), so the value finally returned is success; otherwise, the value of the expression beginning with TAG is that of (justify query program NIL), which will then be failure. The second part of the old invoke, now called writesolution, displays the list of substitutions which have allowed the stack to be emptied; this takes place at the same moment that the stack of goals is found to be empty.

By adding the procedures justify, invoke and writesolution just described to program 3.3 and by getting rid of the previous versions of justify and invoke, we obtain a completely equivalent interpreter, meaning that it stops as soon as the first proof is found. On the other hand, in program 3.4 below some simple new modifications are suggested (they are underlined) which allow the search to be continued when the user so desires. The value of the variable **another** passed to microprolog indicates whether or not the user wants the search to be reinvoked automatically; the value of the variable **repeat**, which is declared in invoke and may be updated by justify, permits the messages issued by invoke and writesolution to be tailored. Just as before, the unification procedure and its utilities are not shown (they remain unchanged: see program 3.1). Session 3.5 shows program 3.4 in operation.

Program 3.4

```
(DE microprolog another ;Main procedure normally called by the user
    (PROG (command)
    (UNTIL NIL
        (prompt "YOUR COMMAND")
        (SETQ command (READ))
        (COND (  (EQ 'S command) (RETURN 'bye))
              (  (EQ '? command)
                 (PRINT "The commands understood are:")
                 (PRINT
                    "NP,OP (New or Old Program), OQ (Old Query),")
                 (PRINT "SP,SQ (Save Program or Query)")
                 (PRINT "E (LISP Evaluation), S (Stop) and ?"))
              (  (EQ 'NP command) (prompt "text of New Program ?")
                 (SETQ program (READ)))
              (  (EQ 'OP command) (prompt "name of Old Program ?")
                 (SETQ program (EVAL (READ))))
              (  (EQ 'OQ command) (prompt "name of Old Query ?")
                 (SETQ query (EVAL (READ))) (invoke))
              (  (EQ 'SP command)
                 (prompt "Save Program by what name ?")
                 (SET (READ) program))
              (  (EQ 'SQ command) (prompt "Save Query by what name ?")
                 (SET (READ) query))
              (  (EQ 'E command) (prompt "Evaluate what expression ?")
                 (PRINT (EVAL (READ))))
              (T (SETQ query command) (invoke))))))

(DMC $ () (LIST '* (READ))) ;macro-character definition

(DF prompt message ;Called by microprolog
    (TERPRI) (PRINT (CAR message) " => "))

(DE invoke () ;Called by microprolog
    (PROG (index repeat)
        (SETQ index 0)
        (COND ( (EQ 'failure (TAG finished (justify query program NIL)))
               (TERPRI)
               (PRINT "I can't find any"
               (COND (repeat " new ") (T " "))
               "justification")))))
```

```
(DE writesolution (allsubst) ;Called by justify
    (COND ((NULL allsubst) (TERPRI) (PRINT "that's correct !"))
          (T (COND (repeat (PRINT "or else"))
                   (T (TERPRI)
                      (PRINT
                         "*** substitutions needed to satisfy your query:")))
             (format (paste (globalsortout allsubst NIL))))))

(DE globalsortout (exlist newlist) ;Called by writesolution
    (COND (  (NULL exlist) (REVERSE newlist))
          (  (NOT (NULL (CDDAAR exlist)))
             (globalsortout (CDR exlist)
                            (applysubstit (LIST (CAR exlist)) newlist)))
          (T (globalsortout (CDR exlist) (CONS (CAR exlist) newlist)))))

(DE paste (expression) ;Called by writesolution
    (COND ((ATOMP expression) expression)
          ((EQ '* (CAR expression))
           (APPLY 'CONCAT (CONS '|$| (CDR expression))))
          (T (MAPCAR (LAMBDA (x) (paste x)) expression))))

(DE format (listsubs) ;Called by invoke
    (PRIN '{)
    (MAPC (LAMBDA (x) (PRIN '| | (CAR x))
                      (PRIN '=) (PRIN (CADR x) '| |)) listsubs)
    (PRINT '}))

(DE justify (goalstack restprog allsubst) ;Called by invoke and justify
    (PROG (resultunif anitemofknowledge)
          (COND ( (NULL goalstack)
                  (writesolution allsubst)
                  (COND ((another?) (SETQ repeat T) (RETURN 'failure))
                        (T (EXIT finished 'success))))
                ( (NULL restprog) (RETURN 'failure)))
          (SETQ resultunif
                (unification (CAR goalstack)
                (CAR (SETQ anitemofknowledge (rename (CAR restprog))))))
          (COND ( (NEQ 'failure resultunif)
                  (justify
                    (applysubstit
                      resultunif (APPEND (CDR anitemofknowledge)
                                         (CDR goalstack)))
                    program
                    (APPEND allsubst resultunif))))
          (justify goalstack (CDR restprog) allsubst)))

(DE another? () ;Called by justify
    (COND (another)
          (T (prompt "another ? (yes or no)") (COND ((EQ 'yes (READ)))))))

(DE rename (e) ;Called by justify
    (SETQ index (1+ index)) (renam1 e))
```

```
(DE renam1 (e) ;Called by rename and renam1
    COND ((ATOM e) e)
           ((EQUAL (CAR e) '*) (LIST '* (CADR e) index))
           ((MAPCAR 'renam1 e))))
```

```
(DE applysubstit (listsubs listexpr) ;Called by justify and globalsortout
    (UNTIL (NULL listsubs)
          (SETQ listexpr (SUBST (CADAR listsubs) (CAAR listsubs) listexpr))
          (SETQ listsubs (CDR listsubs)))
    listexpr)
```

Session 3.5
(microprolog)

YOUR COMMAND =>
NP

text of New Program ? =>
(((fair mark)) ((dark john)) ((father peter john))
  ((father mark peter)) ((father john robert)) ((father $x $y) (son $y $x))
  ((grandfather $x $y) (father $x $z) (father $z $y)) ((son mark george)))

YOUR COMMAND =>
SP

Save Program by what name ? =>
p1

YOUR COMMAND =>
((father $x $y))

*** substitutions needed to satisfy your query:
{ $x=peter $y=john }

another ? (yes or no) =>
yes
or else
{ $x=mark $y=peter }

another ? (yes or no) =>
yes
or else
{ $x=john $y=robert }

another ? (yes or no) =>
yes
or else
{ $x=george $y=mark }

another ? (yes or no) =>
yes
I can't find any new justification

YOUR COMMAND =>
S

= bye

(microprolog t)

YOUR COMMAND =>
E

Evaluate what expression ? =>
p1
(((fair mark)) ((dark john)) ((father peter john)) ((father mark peter))
((father john robert)) ((father (* x) (* y)) (son (* y) (* x))) ((grandfather
(* x) (* y)) (father (* x) (* z)) (father (* z) (* y))) ((son mark george)))

YOUR COMMAND =>
((grandfather $x $y))

*** substitutions needed to satisfy your query:
{ $x=peter $y=robert }
or else
{ $x=mark $y=john }
or else
{ $x=george $y=peter }

I can't find any new justification

YOUR COMMAND =>
S
= bye

The reader could easily make several improvements to program 3.4:

— there could be a specific command for changing the search type (single or repeated solution) between queries without leaving microprolog;
— invocations of the kind (microprolog n) could be allowed, where n is a positive integer, to ask for at most n solutions to be sought;
— queries could be accepted which are not *lists* of goals but rather *sequences* of goals (omission of the outer parentheses).

Session 3.6 tests the microprolog interpreter (program 3.4) on the code of the classic function append, using two different representations:

```
( ((append nil $l $l))
  ((append (ht $u $x) $l (ht $u $z)) (append $x $l $z)) )
```

```
( ((append nil $l $l))
  ((append ($u $x) $l ($u $z)) (append $x $l $z)) )
```

In both cases the first clause (a fact) states that the result of concatenating the empty list and a list $l is the list $l; the second clause (a rule) states that the result of concatenating a list whose head (first element) is $u and whose tail is $x (list of remaining elements) with a list $l is a list whose head is $u and whose tail is $z, where $z is the result of concatenating $x and $l.

For the reader's information, these two versions of append would be written in PROLOG-II as:

```
append(nil,l,l) ->;
append(u.x,l,u.z) -> append(x,l,z);
```

In PROLOG-II, variables are distinguished from constants by the following convention: names of constants start with at least 2 letters. It is also a convention that the full-stop character is the symbol of a function separating the head of a list from its tail: the notation u.x is equivalent to .(u,x), where u is the head of a list whose tail is x.

Note: in the first representation of the program, some function ht (head and tail: this is our own notation, meaning that it is not known to the interpreter in advance) is used explicitly to distinguish the head and tail of a list; in the second representation the auxiliary function ht is only implicit: a list whose first element is a variable, such as ($u $v), stands for a list with head $u and tail $v ($v is not the second element of the **represented** list, but rather its tail). We should point out that any of the 3 arguments of the append predicate can act as the output variable (i.e. can be used to return the result). This property, which arises from constructing results by unification (or, more generally, by *pattern-matching*, see §2.4), is often considered to be one of the attractions of programs written in PROLOG.

Session 3.6
(microprolog t)

YOUR COMMAND =>
NP

text of New Program ? =>
(((append nil $l $l))
  ((append (ht $u $x) $l (ht $u $z)) (append $x $l $z)))

YOUR COMMAND =>
SP

Save Program by what name ? =>
progappend1

YOUR COMMAND =>
((append (ht 1 (ht 3 nil)) (ht 5 (ht 7 (ht 9 nil))) $x))

*** substitutions needed to satisfy your query:
{ $x=(ht 1 (ht 3 (ht 5 (ht 7 (ht 9 nil))))) }

I can't find any new justification

YOUR COMMAND =>
((append $x (ht 5 (ht 7 (ht 9 nil))) (ht 1 (ht 3 (ht 5 (ht 7 (ht 9 nil)))))))

*** substitutions needed to satisfy your query:
{ $x=(ht 1 (ht 3 nil)) }

I can't find any new justification

YOUR COMMAND =>
((append $x $y (ht 1 (ht 3 (ht 5 nil)))))

*** substitutions needed to satisfy your query:
{ $x=nil $y=(ht 1 (ht 3 (ht 5 nil))) }

or else
{ $x=(ht 1 nil) $y=(ht 3 (ht 5 nil)) }
or else
{ $x=(ht 1 (ht 3 nil)) $y=(ht 5 nil) }
or else
{ $x=(ht 1 (ht 3 (ht 5 nil))) $y=nil }

I can't find any new justification

YOUR COMMAND =>
NP

text of New Program ? =>
(((append nil $l $l))
  ((append ($u $x) $l ($u $z)) (append $x $l $z)))

YOUR COMMAND =>
((append (1 (3 nil)) (5 (7 (9 nil))) $x))

∗∗∗ substitutions needed to satisfy your query:
{ $x=(1 (3 (5 (7 (9 nil))))) }

I can't find any new justification

YOUR COMMAND =>
((append $x (5 (7 (9 nil))) (1 (3 (5 (7 (9 nil)))))))

∗∗∗ substitutions needed to satisfy your query:
{ $x=(1 (3 nil)) }

I can't find any new justification

YOUR COMMAND =>
((append $x $y (1 (3 (5 nil)))))

∗∗∗ substitutions needed to satisfy your query:
{ $x=nil $y=(1 (3 (5 nil))) }
or else
{ $x=(1 nil) $y=(3 (5 nil)) }
or else
{ $x=(1 (3 nil)) $y=(5 nil) }
or else
{ $x=(1 (3 (5 nil))) $y=nil }

I can't find any new justification

YOUR COMMAND =>
S
= bye

;in the same session we call up microprolog again,
;but this time without any argument
(microprolog)

YOUR COMMAND =>
OP

name of Old Program ? =>

progappend1

YOUR COMMAND =>
((append $x (ht 1 (ht 3 nil)) $y))

*** substitutions needed to satisfy your query:
{ $x=nil $y=(ht 1 (ht 3 nil)) }

another ? (yes or no) =>
yes
or else
{ $x=(ht $u2 nil) $y=(ht $u2 (ht 1 (ht 3 nil))) }

another ? (yes or no) =>
yes
or else
{ $x=(ht $u2 (ht $u4 nil)) $y=(ht $u2 (ht $u4 (ht 1 (ht 3 nil)))) }

another ? (yes or no) =>
no

YOUR COMMAND =>
S
= bye

3.2.6   Fourth PROLOG interpreter: the control predicate / (cut)
Existing PROLOG dialects possess a predicate for controlling the search
called "cut", often written "/"†. This is a *predefined predicate*, which means
a predicate to which the interpreter attaches a precise built-in interpretation. We shall introduce further predefined predicates later on (§3.2.7).

Here are three equivalent representations of PROLOG programs defining the predicate member which tests whether an element belongs to a list:

> (((member $x (ht $x $y)))
>   ((member $x (ht $y $z)) (member $x $z)))

> (((member $x ($x $y)))
>   ((member $x ($y $z)) (member $x $z)))

> member(x,x.y) ->;
> member(x,y.z) -> member(x,z);

The first two representations are the equivalents of the two ways of writing
the append predicate used in session 3.6. The last representation conforms
to PROLOG-II conventions. Recall that the logical clause set associated
with these programs is:

$$\{MEMBER(X,ht(X,Y)), MEMBER(X,ht(Y,Z)) \lor \square\ MEMBER(X,Z)\}$$

The notation used is that of §3.2; the two-argument function ht is used to
represent lists, the first argument being the head, the second being the tail.
Let us consider a PROLOG interpreter of the "automatic repeated search"

† Trans.: English-speaking readers may be more familiar with the "!" used in Edinburgh
syntax.

type and examine the graph of derivations performed by the interpreter, given one of the preceding PROLOG programs, when it deals with the query: member(3,1.3.2.3.nil). We shall use the PROLOG-II version, the instruction clauses being designated C1 and C2:

$$\text{C1: member(x,x.y) ->;}$$
$$\text{C2: member(x,y.z) -> member(x,z).}$$

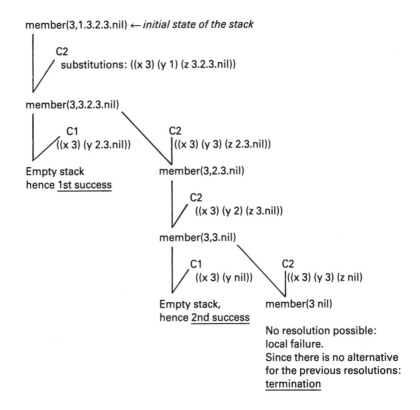

Fig 3.1.

The renaming of variables has not been shown in Fig. 3.1. Resolving the top of the goal stack (here the stack is always limited to one element) with C2 and then C1 allows the stack to be emptied.

After this first success, the interpreter looks for an alternative to the preceding solutions by *chronological* backtracking; in this case, the last resolution (the one which involved C1) is undone; after successive resolutions with C2, C2 a second time and then C1, the stack is once again empty; at the conclusion of this second success the interpreter finds an alternative involving C2, then stops because there are no more alternatives. At the time of the 2 successes the interpreter will issue a message such as: {} since there are no variables in the query. Now consider the following program C'1 and C2:

C1: member(x,x.y) -> /;
C2: member(x,y.z) -> member(x,z);

and follow the working of the interpreter shown in Fig. 3.2.

member(3,1.3.2.3.nil) ← *initial state of the stack*

C2
substitutions: ((x 3) (y 1) (z 3.2.3.nil))

member(3,3.2.3.nil)

C1   C2
((x 3) (y 2.3.nil))

When a / goal comes to the top of the stack, the action is as though there were a fact: / ->; and no other instruction-clause beginning with /. Thus the goal / is immediately eliminated and no alternative way of resolving the / exists. Furthermore, all of the possible alternative resolutions of the goal whose elimination introduced the /, and all of the goals considered up to the point where / appeared at the top of the stack, are removed; hence / "cuts" potential branches in the graph of the resolvents that could be produced from the query.

Empty stack, hence success.
Because of the cuts due to "resolving" /, no alternative remains, hence termination.

Fig 3.2.

Let us consider a more general example of the use of /, which also serves as a supporting explanation in (FARRENY & GHALLAB 1986); the program is written in PROLOG-II style; the letters A, B, etc. stand for symbols of any distinct predicates (there are no variables to complicate the issue here).

A -> B C D E ;
A -> ...
...   In this example,
B -> ;   the dots indicate that there are
...   groups of facts and rules with
C -> ;   the predicate symbol as the head
...
D -> F G / H J ;

...
F -> ;   Note: there is no fact or rule with J as the
                                    head

...
G -> ;
...
H -> ;
H -> I ;
E -> ;
...

The various possible resolutions can be represented by an AND-OR graph diagram (see Fig. 3.3). In this graph, the node symbolised by D is the

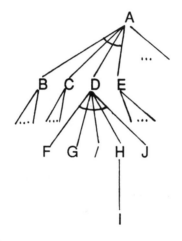

Fig 3.3.

father-node of *I*: it is when the occurrence of predicate D represented by this particular node is resolved that *I* is introduced. Nodes F, G, H and J are the *sibling-nodes* of *I*. F and G are its *elder* sibling-nodes, H and J are its *younger* sibling-nodes. Consider how goal A is treated; the sub-goals B and C are resolved straightaway, then D is decomposed into F, G, *I*, H and J. Sub-goals F, G, *I* and H are resolved immediately while the resolution of J fails. The previous resolution of H is undone.

The rule H ->I is invoked, but resolving I proves impossible. There is no further alternative for resolving H. Had the *I* predicate not been present, the interpreter could have investigated the other possibilities for satisfying G, then F, then D. Now, the role of *I*, once it is pulled off the goal stack, is precisely to reject all other possible resolutions of the elder sibling-nodes *and* the father-sibling that had yet to be considered (the father-node of a *I* appearing in the initial query is undefined). The *cuts* imposed by resolving the / goal are shown in the preceding arborescence.

In order to understand the action of / properly and thereby be able to appreciate what is happening later on (program 3.5), let us take an example in which there are several occurrences of the / predicate (they are marked below as an aid to the discussion):

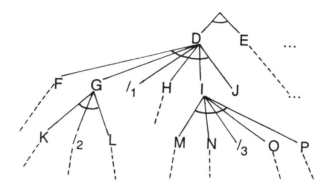

Fig. 3.4.

Suppose that the current goal, D, is broken down into sub-goals F, G, /, H, I and J. Assume that F can be satisfied, possibly after several attempts and therefore several backtracks. Suppose that G is broken down into K, / and L. Then K is resolved, again possibly after several tries. By the definition of /, $/_2$ succeeds immediately. Eliminating L can give rise to many alternatives. If it fails, the interpreter must (as a result of having satisfied $/_2$) return directly to resolving F (without retrying K or G). If there is a new way of solving F, all possible ways of solving G will be reconsidered. If G is eventually satisfied, $/_1$ immediately succeeds. Assume that H is then resolved, perhaps after several attempts. Suppose also that a rule allows I to be decomposed into M, N, $/_3$, O and P.

If M and N are satisfied, possibly after several tries, $/_3$ succeeds; but should the sequence of goals O, P fail, the interpreter (because of $/_3$) must not investigate other means of resolving N, M or I: it must go straight back to re-examining the resolution of H. If an alternative for H is found, all ways of satisfying I will be considered or reconsidered. If satisfying the sequence H, I, J finally fails, the interpreter must not retry L, K, G, F or D (because of $/_1$), unless it can find a new means of a solving a goal which had been satisfied prior to the resolution of D at which we began here. If the resolution of the series H, I, J succeeds but the resolution of E fails (in this case E is the younger sibling of D, but more generally speaking E could be a goal which is already on the goal stack when D arrived at the top), the interpreter must not try fresh solutions for D, since the alternatives for D were discarded at the time $/_1$ was satisfied (once again unless a new solution to a goal satisfied before D is found).

We can see that for each occurrence of / appearing on the stack of goals there is a *set of goals with alternatives controlled by the cut*, whose elements are the cut's father-node (if it exists), the elder-nodes and their descendants (like K and L in relation to $/_1$), and a *set of goals liable to activate the cut*, whose elements

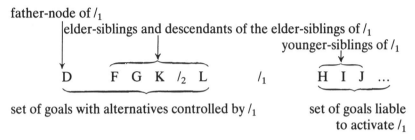

Fig. 3.5

are all of the goals on the goal stack lying after the cut (starting with younger sibling-nodes of the /). If all the alternatives for satisfying one of the goals in the latter set are exhausted, the cut is *activated* in the sense that the interpreter has to ignore the alternatives available for each of the goals in the former set. In the case of the previous example, if $/_1$ comes to the top of the stack after D, F, G, K, $/_2$ and L have been resolved, the situation represented in Fig. 3.5 is obtained.

Note that as a result of the strategy for satisfying goals which is followed by PROLOG interpreters, if two occurrences of / are simultaneously present on the goal stack, the set of goals on the stack whose alternatives are controlled by the / which is the furthest away from the top of the stack includes the set of goals on the stack whose alternatives are controlled by the other /.

In order to equip PROLOG interpreter 3.4 with the capacity to handle cut predicates, each time a cut occurs it must be able to recognise and process the two associated goal sets. This is what the new version of **justify** presented below accomplishes.

Alterations to the previous version are underlined.

Let us examine the sequence of goals that is passed to a call of **justify**. This call, and the corresponding occurrence in the code of **justify**, will be termed CC: Current Call. If the value passed to the parameter **goalstack** or the parameter **restprog** (line 1 of **justify**) is the empty list, **justify** behaves as in the previous version: either a solution is found and the search is not set off again (CC is abandoned via (**EXIT finished 'success**)), or else CC returns an indication of failure (in this case: **failurewithoutcut**, see lines 3 to 7). Let B be the first element of **goalstack** at the point when line 8 is reached. Lines 8 to 11 are intended to cope with the situation where B is a cut-type goal; we shall comment on these lines later on.

When line 12 is executed, B is an ordinary goal (no cut). As in the old version, a call is made to unify B and the head of the first clause in the PROLOG program (**restprog**), the result being assigned to **resultunif**.

```
1   (DE justify (goalstack restprog allsubst) ;Called by invoke and justify
2   (PROG (resultunif resultjustif anitemofknowledge cutintroduced)
3     (COND (  (NULL goalstack)
4                (writesolution allsubst)
5                (COND ((another?) (SETQ repeat T) (RETURN' failurewithoutcut))
6                      (T (EXIT finished 'success))))
7           (  (NULL restprog) (RETURN 'failurewithoutcut))
8           (  (AND (LISTP (CAR goalstack)) (EQ (CAAR goalstack) '/))
9                (SETQ resultjustif (justify (CDR goalstack) program allsubst))
10               (COND ((EQ 'failurewithoutcut resultjustif)
                           (RETURN (CADAR goalstack)))
11                     (T (RETURN resultjustif))))
12          (T (SETQ resultunif
13                (unification
14                  (CAR goalstack)
15                  (CAR (SETQ anitemofknowledge (rename (CAR restprog))))))
16             (COND ( (NEQ 'failure resultunif)
17                       (COND ( (MEMQ '/ (CDR anitemofknowledge))
18                                 (SETQ cutintroduced (GENSYM))))
19                       (SETQ
20                         resultjustif
21                         (justify
22                           (applysubstit
23                           resultunif
24                           (APPEND
25                             (SUBST (LIST '/ cutintroduced) '/
                                     (CDR anitemofknowledge))
26                           (CDR goalstack)))
27                         program
28                         (APPEND allsubst resultunif)))))))
29    (COND (  (OR (EQ 'failure resultunif)
                     (EQ 'failurewithoutcut resultjustif))
30               (justify goalstack (CDR restprog) allsubst))
31            ((EQ cutintroduced resultjustif) (RETURN 'failurewithoutcut))
32            (T (RETURN resultjustif)))))
```

If the unification succeeds, justify is invoked, as in the previous version, starting with the stack of goals passed to CC but with the top of B replaced by the tail of the first clause, modulo the substitutions carried out by the unification. The tail of the first clause introduces child-goals of B which may include / symbols. In order to identify for each occurrence of / the *set of goals with alternatives controlled by this* / and *the set of goals liable to activate this* / (see the definitions of these sets above), a unique tag is created by calling GENSYM (line 18; call this tag G_i) and assigned to the local variable cutintroduced (whose value is otherwise the empty list); then (line 25) each / that is a child of B is replaced by a list of the form: $(/ G_i)$. The result of justify is set to resultjustif.

If the goal $(/ Gi)$ comes to the top of the stack later on (line 8), it is considered to be immediately resolved since a new call of justify is made straightaway (line 9) with parameters: the stack of goals without the goal (G_i), the whole PROLOG program and the set of already stored substitutions. The result of justify is again set to resultjustif in this call. If it manages to empty (CDR goalstack) and the search is not to be reinvoked, an escape similar to the one in line 6 will be taken, hence no return to the calling procedure CC will be made; otherwise, as can be seen by examining the reorders in lines 5, 7, 10, 11, 30, 31 and 32, the justify call can only return either the tag failurewithoutcut or a tag of type G_j. First case: if the call to justify returns failurewithoutcut, CC returns the value G_i (end of line 10) to indicate that the occurrence of cut located by G_i is activated from now on. Second case: whenever the call to justify does not return failurewithoutcut it returns a resultjustif value equal to G_j to signify that an occurrence of cut located by G_j has already been and still is activated. It is this value G_j that must be returned (line 11) since, as we pointed out earlier, if two occurrences of / are present on the stack of goals at the same time, the alternatives controlled by the occurrence furthest from the top (here the one by G_j) include those controlled by the other (here found by G_i).

Line 30 allows a third call of justify with the same stack of goals (goalstack) as that passed to CC and with the same set of stored substitutions (allsubst) as CC too, but with the program restprog minus its first clause. When the interpreter reaches the tests in line 29 which decide whether line 30 is to be executed, one of two mutually exclusive situations will have been encountered:

— either the unification of lines 13 to 15 between B and the head of the first clause has failed, in which case, as in version 3.4, justify must be reinvoked with the same goal stack and the same set of substitutions, but with the program minus one clause (line 30);
— or else the unification of lines 13 to 15 has succeeded, in which case lines 19 to 28 have been executed (justify has been called);
  > if resultjustif is set to failurewithoutcut then, as in the previous case and in version 3.4, justify has to be reinvoked with the same goal stack and the same set of substitutions, but with the program minus one clause (line 30);
  > if on the other hand resultjustif is not set to failurewithoutcut, i.e. it is set to a G_j tag (remember that from justify can only be failurewithoutcut or a G_j type tag signifying that a particular occurrence of / is active), no alternative way of satisfying B must be sought: since the call of justify in line 21 has returned G_j, B is one of the goals whose alternatives are controlled by the / denoted by G_j; had this not been so, line 21's call of justify would have returned failurewithoutcut. In fact, line 31 checks whether the G_j tag received by resultjustif is equal to the value of cutintroduced; if this is the case, the goal B is recognised as the father-node of the / denoted by G_j. Hence B is the last goal whose alternatives are controlled by this /: CC must now return failurewithoutcut and not G_j (otherwise the father-goal of the / has not yet been found, so B is one of the goals whose alternatives are

controlled by this cut, the cut will remain active, and G_j will be returned: line 32).

The enhancements to justify which have just been described permit us to handle occurrences of / introduced in the PROLOG program being interpreted. In order to cope with those appearing in queries, the invoke procedure has only to be modified in analogous terms to those of lines 17, 18 and 25 of justify (the underlined section is new):

```
(DE invoke () ;Called by microprolog
    (PROG (index repeat)
        (SETQ index 0)
        (COND ( (MEMQ '/ query)
                (SETQ query (SUBST (LIST '/ (GENSYM)) '/ query))))
        (COND ( (NEQ 'success (TAG finished (justify query program NIL)))
                (TERPRI)
                (PRINT "I can't find any"
                    (COND (repeat " new ") (T " "))
                    "justification")))))
```

Program 3.5 below only differs from program 3.4 in its justify and invoke procedures. As before, the unification procedure and its utilities are not shown (program 3.1).

Program 3.5

```
(DE microprolog another ;Main procedure normally called by the user
    (PROG (command)
    (UNTIL NIL
        (prompt "YOUR COMMAND")
        (SETQ command (READ))
        (COND ( (EQ 'S command) (RETURN 'bye))
               ( (EQ '? command)
                 (PRINT "The commands understood are:")
                 (PRINT
                    "NP,OP (New or Old Program), OQ (Old Query),")
                 (PRINT "SP,SQ (Save Program or Query)")
                 (PRINT "E (LISP Evaluation), S (Stop) and ?"))
               ( (EQ 'NP command) (prompt "text of New Program ?")
                 (SETQ program (READ)))
               ( (EQ 'OP command) (prompt "name of Old Program ?")
                 (SETQ program (EVAL (READ))))
               ( (EQ 'OQ command) (prompt "name of Old Query ?")
                 (SETQ query (EVAL (READ))) (invoke))
               ( (EQ 'SP command) (prompt
                 "Save Program by what name ?")
                 (SET (READ) program))
               ( (EQ 'SQ command) (prompt "Save Query by what name ?")
                 (SET (READ) query))
               ( (EQ 'E command) (prompt "Evaluate what expression ?")
                 (PRINT (EVAL (READ))))
               (T (SETQ query command) (invoke))))))
```

```
(DMC $ () (LIST '* (READ))) ;macro-character definition

(DF prompt message ;Called by microprolog
    (TERPRI) (PRINT (CAR message) " => "))

(DE invoke () ;Called by microprolog
    (PROG (index repeat)
        (SETQ index 0)
        (COND ( (MEMQ '/ query)
                (SETQ query (SUBST (LIST '/ (GENSYM)) '/ query))))
        (COND ( (NEQ 'success (TAG finished (justify query program NIL)))
                (TERPRI)
                (PRINT "I can't find any"
                    (COND (repeat " new ") (T " "))
                    "justification")))))

(DE writesolution (allsubst) ;Called by justify
    (COND ((NULL allsubst) (TERPRI) (PRINT "that's correct !"))
          (T (COND (repeat (PRINT "or else"))
                   (T (TERPRI)
                      (PRINT
                          "*** substitutions needed to satisfy your query:")))
             (format (paste (globalsortout allsubst NIL))))))

(DE format (listsubs) ;Called by invoke
    (PRIN '{)
    (MAPC (LAMBDA (x) (PRIN '| | (CAR x)) (PRIN '=) (PRIN (CADR x) '| |))
          listsubs)
    (PRINT '}))

(DE justify (goalstack restprog allsubst) ;Called by invoke and justify
    (PROG (resultunif resultjustif anitemofknowledge cutintroduced)
        (COND (   (NULL goalstack)
                  (writesolution allsubst)
                  (COND ((another?) (SETQ repeat T)
                                    (RETURN 'failurewithoutcut))
                        (T (EXIT finished 'success))))
              (   (NULL restprog) (RETURN 'failurewithoutcut))
              (   (AND (LISTP (CAR goalstack)) (EQ (CAAR goalstack) '/))
                  (SETQ resultjustif (justify (CDR goalstack) program allsubst))
                  (COND ((EQ 'failurewithoutcut resultjustif)
                         (RETURN (CADAR goalstack)))
                        (T (RETURN resultjustif))))
              (T (SETQ resultunif
                       (unification
                         (CAR goalstack)
                         (CAR (SETQ anitemofknowledge
                                    (rename (CAR restprog))))))
                 (COND ( (NEQ 'failure resultunif)
                         (COND ( (MEMQ '/ (CDR anitemofknowledge))
                                 (SETQ cutintroduced (GENSYM))))
                         (SETQ
```

```
                            (resultjustif
                            (justify
                              (applysubstit
                                resultunif
                                (APPEND
                                (SUBST
                                  (LIST '/ cutintroduced) '/
                                  (CDR anitemofknowledge))
                                  (CDR goalstack)))
                              program
                              (APPEND allsubst resultunif)))))))
              (COND (   (OR (EQ 'failure resultunif) (EQ 'failurewithoutcut
                resultjustif)) (justify goalstack (CDR restprog) allsubst))
                        (EQ cutintroduced resultjustif) (RETURN 'failurewithoutcut))
                (T (RETURN resultjustif)))))

(DE another? () ;Called by justify
    (COND (another)
          (T (prompt "another ? (yes or no)") (COND ((EQ 'yes (READ)))))))

(DE globalsortout (exlist newlist) ;Called by writesolution
    (COND (   (NULL exlist) (REVERSE newlist))
          (   (NOT (NULL (CDDAAR exlist)))
              (globalsortout (CDR exlist)
                      (applysubstit (LIST (CAR exlist)) newlist)))
          (T (globalsortout (CDR exlist) (CONS (CAR exlist) newlist)))))

(DE paste (expression) ;Called by writesolution
    (COND ((ATOMP expression) expression)
          ((EQ '* (CAR expression))
           (APPLY 'CONCAT (CONS '|$| (CDR expression))))
          (T (MAPCAR (LAMBDA (x) (paste x)) expression))))

(DE rename (e) ;Called by justify
    (SETQ index (1+ index)) (renam1 e))

(DE renam1 (e) ;Called by rename and renam1
    (COND ((ATOM e) e)
          ((EQUAL (CAR e) '*) (LIST '* (CADR e) index))
          ((MAPCAR 'renam1 e))))

(DE applysubstit (listsubs listexpr) ;Called by justify and globalsortout
    (UNTIL (NULL listsubs)
        (SETQ listexpr (SUBST (CADAR listsubs) (CAAR listsubs) listexpr))
        (SETQ listsubs (CDR listsubs)))
    listexpr)
```

Session 3.7 demonstrates the behaviour of interpreter 3.5. The second PROLOG program is different from the one saved as **p1** in that it has a / in its first clause. The presence of / in the second program means that the interpreter looks for only one justification for queries **q1** and **q2**, while the

way **p1** is written does not stop the search from being reinvoked. The third
and fourth queries to **p1** each contain a /; even when compared against **p1**,
they limit the search to a single justification.

Session 3.7
(microprolog t)

YOUR COMMAND =>
NP

text of New Program ? =>
(((member $x (ht $x $y))) ((member $x (ht $y $z)) (member $x $z)))

YOUR COMMAND =>
SP

Save Program by what name ? =>
p1

YOUR COMMAND =>
((member 3 (ht 1 (ht 3 (ht 2 (ht 3 nil))))))

∗∗∗ substitutions needed to satisfy your query:
{}
or else
{}

I can't find any new justification

YOUR COMMAND =>
SQ

Save Query by what name ? =>
q1

YOUR COMMAND =>
((member $x (ht 1 (ht 3 (ht (ht 2 (ht 4 nil)) (ht 3 nil))))))

∗∗∗ substitutions needed to satisfy your query:
{ $x=1 }
or else
{ $x=3 }
or else
{ $x=(ht 2 (ht 4 nil)) }
or else
{ $x=3 }

I can't find any new justification

YOUR COMMAND =>
SQ

Save Query by what name ? =>
q2

YOUR COMMAND =>
((member 3 (ht 1 (ht 3 (ht 2 (ht 3 nil))))) /)

*** substitutions needed to satisfy your query:
{}

I can't find any new justification

YOUR COMMAND =>
((member $x (ht 1 (ht 3 (ht (ht 2 (ht 4 nil)) (ht 3 nil)))))) /)

*** substitutions needed to satisfy your query:
{ $x=1 }

I can't find any new justification

YOUR COMMAND =>
NP

text of New Program ? =>
(((member $x (ht $x $y)) /) ((member $x (ht $y $z)) (member $x $z)))

YOUR COMMAND =>
OQ

name of Old Query ? =>
q1

*** substitutions needed to satisfy your query:
{}

I can't find any new justification

YOUR COMMAND =>
OQ

name of Old Query ? =>
q2

*** substitutions needed to satisfy your query:
{ $x=1 }

I can't find any new justification

YOUR COMMAND =>
S
= bye

Session 3.8 handles a purely symbolic example of the kind discussed in connection with Fig. 3.4. In this example the goal stack can contain multiple occurrences of / at a given instant. To help the reader to follow what **justify** is doing and verify that it conforms to the preceding comments, several commands for printing messages have been inserted into the body of **justify** (underlined sections below); apart from these minor alterations, the interpreter used in the session remains the same as that described by program 3.5.

```
(DE justify (goalstack restprog allsubst) ;Called by invoke and justify
  (PROG (resultunif resultjustif anitemofknowledge cutintroduced)
    (COND (  (NULL goalstack)
             (writesolution allsubst)
             (COND ((another?) (SETQ repeat T) (RETURN
                       'failurewithoutcut))
                   (T (EXIT finished 'success))))
          (  (NULL restprog)
             (TERPRI)
             (PRINT "program empty with goalstack equal to: " goalstack)
             (RETURN 'failurewithoutcut))
          (  (AND (LISTP (CAR goalstack)) (EQ (CAAR goalstack) '/))
             (SETQ resultjustif (justify (CDR goalstack) program allsubst))
             (COND (  (EQ 'failurewithoutcut resultjustif)
                      (TERPRI) (PRINT "goalstack equal to: " goalstack)
                      (PRINT "activation of the cut with tag: "
                      (PRINT (CADAR goalstack))
                      (RETURN (CADAR goalstack)))
                   (T (TERPRI) (PRINT "goalstack equal to: " goalstack)
                      (PRINT "the result is simply: " resultjustif)
                      (RETURN resultjustif))))
          (T (SETQ resultunif
                  (unification
                    (CAR goalstack)
                    (CAR (SETQ anitemofknowledge
                               (rename (CAR restprog))))))
             (COND ( (NEQ 'failure resultunif)
                     (TERPRI) (PRINT "goalstack equal to: " goalstack)
                     (PRINT "unification succeeded with head of: "
                            anitemofknowledge)
                     (COND ( (MEMQ '/ (CDR anitemofknowledge))
                             (SETQ cutintroduced (GENSYM))))
                     (SETQ
                       resultjustif
                       (justify
                         (applysubstit
                           resultunif
                           (APPEND
                           (SUBST
                            (LIST '/ cutintroduced) '/
                            (CDR anitemofknowledge))
                            (CDR goalstack)))
                         program
                         (APPEND allsubst resultunif)))))))
    (COND (  (OR (EQ 'failure resultunif)
                 (EQ 'failurewithoutcutresultjustif))
             (justify goalstack (CDR restprog) allsubst))
          (  (EQ cutintroduced resultjustif)
             (TERPRI) (PRINT "goalstack equal to: " goalstack)
             (PRINT deactivation of the cut with tag: " resultjustif)
             (RETURN 'failurewithoutcut))
          (T (TERPRI) (PRINT "goalstack equal to: " goalstack)
             (PRINT "the result is simply: " resultjustif)
             (RETURN resultjustif)))))
```

Session 3.8:
(microprolog t)

YOUR COMMAND =>
NP

text of New Program ? =>
((a d e)
  (e)
  (d f g / h i j) (d)
  (f)
  (g k / l) (g f) (g)
  (k) (k f)
  (h)
  (i m n / o p) (i)
  (m) (m f)
  (n) (n f)
  (o))

YOUR COMMAND =>
(a)

goalstack equal to: (a)
unification succeeded with head of: (a d e)

goalstack equal to: (d e)
unification succeeded with head of: (d f g / h i j)

goalstack equal to: (f g (/ g104) h i j e)
unification succeeded with head of: (f)

goalstack equal to: (g (/ g104) h i j e)
unification succeeded with head of: (g k / l)

goalstack equal to: (k (/ g105) l (/ g104) h i j e)
unification succeeded with head of: (k)

program empty with goalstack equal to: (l (/ g104) h i j e)

goalstack equal to: ((/ g105) l (/ g104) h i j e)
activation of the cut with tag: g105

goalstack equal to: (k (/ g105) l (/ g104) h i j e)
the result is simply: g105

goalstack equal to: (g (/ g104) h i j e)
deactivation of the cut with tag: g105

program empty with goalstack equal to: (f g (/ g104) h i j e)

goalstack equal to: (d e)
unification succeeded with head of: (d)

goalstack equal to: (e)
unification succeeded with head of: (e)

that's correct !

program empty with goalstack equal to: (e)

program empty with goalstack equal to: (d e)

program empty with goalstack equal to: (a)

I can't find any new justification

YOUR COMMAND =>
S
= bye

3.2.7 Fifth PROLOG interpreter: predefined predicates, evaluable functions

The / predicate was the first example of a *predefined predicate*. The user does not have to write clauses defining the /, unlike what he has to do when he wishes to declare his own predicates (for instance member). When a / comes to the top of the goal stack, it is, by definition, immediately 'resolved', hence immediately eliminated *without looking at the clauses in the user's PROLOG program.*

The / is a particular kind of goal such that (i) the 'resolution' is deemed to have succeeded without trying any unification with the head of the first clause in the user's program (the interpreter no longer follows the basic strategy of applying the resolution principle), (ii) the resolution does not introduce sub-goals onto the stack, (iii) there is no alternative resolution to the first one, (iv) the elimination of the goal from the stack is accompanied by *side-effects* which are described in advance in the actual code of the interpreter (and not in the PROLOG program written by the user).

Each PROLOG dialect offers a range of predicates obeying conventions (i), (ii), (iii) and (iv) above. These predicates are called *predefined*. For some of them, such as /, 'resolution' always succeeds by definition; for others it depends on specific tests (thus convention (i) does not automatically mean that the 'resolution' succeeds). The version of justify presented below, which comes from that integrated into program 3.5, allows for an (extensible) family of predefined predicates to be dealt with, in addition to /.

Changes are underlined. Predefined predicates (other than / which is treated individually) are detected by predefined?. The lines which investigate predefined predicates (including /) are now placed before the test: (NULL restprog). In fact, a microprolog session may be executed without any program having been written if only predefined predicates are used. To prevent the Le-Lisp interpreter from complaining ('undefined variable') when invoke uses program, from now on program will be declared as a local variable in microprolog (as the second variable of the PROG after command, automatically initialised to NIL); the restprog parameter to justify will now be passed the empty list at the time of the call, but this does not signify a failure. This is why the test (NULL restprog) has been moved. This problem has not arisen so far for / since, when justify was called with the first element of goalstack being a /, the argument restprog was necessarily not empty: it was the value of program.

```
(DE justify (goalstack restprog allsubst) ;Called by invoke and justify
   (PROG (result resultunif resultjustif anitemofknowledge cutintroduced)
      (COND ( (NULL goalstack)
               (writesolution allsubst)
               (COND ((another?) (SETQ repeat T)
                                 (RETURN 'failurewithoutcut))
                     (T (EXIT finished 'success))))
            ( (AND (LISTP (CAR goalstack)) (EQ (CAAR goalstack) '/))
               (SETQ resultjustif (justify (CDR goalstack) program allsubst))
               (COND ((EQ 'failurewithoutcut resultjustif)
                        (RETURN (CADAR goalstack)))
                     (T (RETURN resultjustif))))
            ( (predefined? (CAR goalstack))
               (SETQ result (COND (  (ATOMP (CAR goalstack))
                                       (EVAL (LIST (CAR goalstack))))
                                  (T (EVAL (CAR goalstack)))))
               (COND ((EQ 'failure result) (RETURN 'failurewithoutcut)))
               (RETURN (justify  (applysubstit result (CDR goalstack))
                           program
                           (APPEND allsubst result))))
            ( (NULL restprog) (RETURN 'failurewithoutcut)))
   (SETQ result (unification
                  (CAR goalstack)
                  (CAR (SETQ anitemofknowledge
                              (rename (CAR restprog))))))
   (COND ( (NEQ 'failure result)
            (COND ( (MEMQ '/ (CDR anitemofknowledge))
                     (SETQ cutintroduced (GENSYM))))
            (SETQ resultjustif
               (justify
                  (applysubstit
                     result
                     (APPEND
                     (topsubst/ cutintroduced)
                     (CDR anitemofknowledge))
                     (CDR goalstack)))
                  program
                  (APPEND allsubst result)))))
   (COND ( (OR (EQ 'failure result) (EQ 'failurewithoutcut resultjustif))
            (justify goalstack (CDR restprog) allsubst))
         (  (EQ cutintroduced resultjustif) (RETURN 'failurewithoutcut))
         (T (RETURN resultjustif))))

(DE predefined? (goal) ;Called by justify
   (COND ((ATOMP goal) (MEMQ goal predefineds))
         (T (MEMQ (CAR goal) predefineds))))
```

For the purpose of illustration, we now define the new predicates: **space**, list, write, assert and lispval. By modelling these 5 examples, the reader-will be able to predefine other commonly used predicates. **predefineds** will be set to the list of predefined predicate symbols:

```
(SETQ predefineds '(space list write assert val))
```

In order to recognise predefined predicates, an alternative technique would be to assign an agreed tag to every atom representing a predefined predicate (but we should then have to be prepared for the risk of a mistake when testing the value of a symbol which does not have a tag). A predefined predicate might also be recognised by testing whether it has a LISP procedure definition attached to it, for example by using **TYPEFN, VALFN** or **GETDEF** in Le-Lisp (but there is a risk of unintentional interference between the user's symbols and those of LISP functions which he does not know about).

Literals using predefined predicates will normally be written in functional form:

(<predefined predicate symbol> <series of arguments>)

For example: (write "call of factorial " $n)
            (assert ((fib $n $r) /))
            (lispval (- $n 2) $temp)

However predicates without arguments can also be predefined in the form of simple identifiers: <predefined predicate symbol>. For instance: **space list**. Such predicates may equally be used in functional form: **(space) (list)**.

The procedure which predefines a predicate has the same name as that predicate. It must return **failure** if the predicate cannot be 'resolved', or else a list of substitutions (which may be empty). Let us illustrate this principle by looking at **lispval**:

```
(DF lispval (arg1 arg2)
    (COND ((variable? arg2) (LIST (LIST arg2 (EVAL arg1))))
          ((EQUAL arg2 (EVAL arg1)) NIL)
          (T 'failure)))
```

This predicate has a similar (but broader) meaning to that of **val** in PROLOG-II. When the **lispval** predicate appears as a goal, the associated procedure, also called **lispval**, is invoked; the first argument is evaluated as a LISP expression; as in PROLOG-II, the function symbols permitted in this first argument must be symbols of *evaluable functions*. PROLOG-II allows: **add, mul, inf, ...**; here any LISP function is permissible (or just about any function: see below) in the text of the first argument. If the second argument is an as yet uninstantiated variable, the value of the first argument is kept as the term part of a substitution dealing with the variable which is the second argument; the **lispval** predicate is taken to be 'resolved' and the result of this pseudo-resolution is a list whose sole element is the above substitution (just such a list was obtained when proper resolutions were carried out); this list will then be treated exactly the same as those resulting from real unifications. Otherwise, the value of the second argument (as it stands, without any LISP evaluation) is compared with that of the first argument (which does come from a LISP evaluation); if the two values are identical (in the sense of **EQUAL** the 'resolution' of **lispval** succeeds and produces an empty list of substitutions; otherwise the 'resolution' fails.

The use of the LISP function ∗ (multiplication) is not allowed in the first argument to lispval: if required, the function times can be used instead. It was agreed in §3.2.4.3 that the ∗ symbol at the head of a list would stand for the internal representation of variables, which renam1 deals with blindly. This convention could of course be reconsidered.

Examples of the use of lispval, and other predefined predicates, are presented in sessions 3.9 and 3.10.

The predicate space (no argument) causes a new line to be thrown, like line in PROLOG-II. It 'succeeds' unconditionally.

(DE space () (TERPRI) NIL)

The predicate list (no argument), just like list in PROLOG-II, causes the list which forms the current PROLOG program to be written out. It 'succeeds' unconditionally. The procedure paste (already used in the previous programs) displays variables in a more compact form.

(DE list () (MAPC 'PRINT (paste program)) NIL)

The write predicate behaves in a similar fashion to out and outm in PROLOG-II. It is 'resolved' unconditionally. Typical examples of the use of write are presented in session 3.10.

(DF write arguments (APPLY 'PRINT (paste arguments)) NIL)

The predicate assert is close to that of the same name in PROLOG-II. It 'succeeds' unconditionally. At the moment it is removed from the top of the stack, its argument is taken to represent the text of a clause which is then placed at the head of the current PROLOG program.

(DF assert (clause) (SETQ program (CONS clause program)) NIL)

As shown in session 3.9, it can be fruitful to modify a previously constructed program *dynamically* (i.e. while it is being interpreted). In the example, as elements of the Fibonacci series are calculated they are stored, using the assert predicate, as PROLOG clauses at the head of the program; the objective is to simplify the determination of new elements by reusing those already calculated rather than recomputing them; in practice, the clauses which are added are *facts* (meaning clauses without a tail) if the presence of / is ignored. Cuts are intended to prevent goals resolved with facts from later on being the object of resolutions with the only *rule* type clause (clause with a tail). The possibility of / occurring in the argument of assert predicates leads us to re-examine the principle of substitution that has been used up to this point: when new goals are introduced onto the stack, all occurrences of / in the new goals were systematically replaced by lists of type (/ G_i) by using SUBST (see justify, program 3.5); now, a / appearing *as a term* of an assert predicate is not a goal (it is not a *predicate*); for this / to be considered as a goal we must wait for (i) the assert goal to succeed and thus a new program clause to be generated (in whose tail the / will appear as a *predicate*), and then (ii) this clause to be resolved and hence the / to be put onto the goal stack (now as a *predicate*, at the same level as the other goals). In fact, the

SUBST procedure was uselessly over-powerful since / *predicates* could only be encountered at the top level of the list of new goals; now that /s can appear as *terms* the procedure SUBST is no longer suitable, so it is replaced by TOPSUBST/ defined below.

```
(DE topsubst/ (code goals) ;Called by justify
    (MAPCAR '(LAMBDA (x) (COND ((EQ '/ code)) (T x))) goals))
```

Session 3.9 was obtained after loading program 3.5 (accompanied by the unification procedures of program 3.1) with the integration of the last version of justify, the procedures predefined?, topsubst/, lispval, space, list, write and assert, the assignment of the global variable predefineds and the declaration of program as a local variable in microprolog.

It should be remembered that for operational use a few program improvements are desirable. In our opinion, there must be a distinction between at least two levels of programming: at the purely functional level (the one considered in this book), problem-solving methods are implemented as straightforwardly and readably as possible; at the operational level, it is the conciseness of the code, its protection vis-à-vis the user's other programs and its speed of execution which are of interest.

The first queries illustrate the role of lispval; note that this predicate's second argument is not evaluated (by the LISP interpreter underlying the PROLOG interpreter), unlike its first argument. The first program represents the calculation of the factorial of an integer; the factorial of 1 is 1, that of $n is $r if $n-1 is $temp1 and if the factorial of $temp1 is $temp2 and if the product of $n and $temp2 is $r; this program makes use of lispval. The second program is similar to the first: in addition it uses space and write. The third program represents the computation of the Fibonacci series whose general term is:

$$un = u_{n-1} + u_{n-2} \text{ with } u_0 = 0 \text{ and } u_1 = 1$$

Session 3.9
(microprolog t)

YOUR COMMAND =>
((lispval (+ 15 5) 20))

that's correct !

I can't find any new justification
YOUR COMMAND =>
((lispval (+ 5 6 7) 10))

I can't find any justification

YOUR COMMAND =>
((lispval (+ 5 6 7) $z))

∗∗∗ substitutions needed to satisfy your query:
{ $z=18 }

I can't find any new justification

YOUR COMMAND =>
((lispval (car (cdr '(a b c))) 'b))

I can't find any justification

YOUR COMMAND =>
((lispval (car (cdr '(a b c))) b))

that's correct !

I can't find any new justification

YOUR COMMAND =>
NP

text of New Program ? =>
(((fact 1 1)
   /)
  ((fact $n $r)
    ((lispval (1- $n) $temp1) (fact $temp1 $temp2) (lispval (times $n $temp2)
    $r)))

YOUR COMMAND =>
((fact 6 $x))

*** substitutions needed to satisfy your query:
{ $x=720 }

I can't find any new justification

YOUR COMMAND =>
NP

text of New Program ? =>
(((fact 1 1)
   /)
  ((fact $n $r)
    space (write "call of factorial " $n)
    (lispval (1- $n) $temp1) (fact $temp1 $temp2) (lispval (times $n $temp2)
    $r)))

YOUR COMMAND =>
((fact 4 $y))

call of factorial 4

call of factorial 3

call of factorial 2

*** substitutions needed to satisfy your query:
{ $y=24 }

I can't find any new justification

```
YOUR COMMAND =>
NP
```

```
text of New Program ? =>
( ( (fib 0 0)
    /)
  ( (fib 1 1)
    /)
  ( (fib $n $r)
    (lispval (1- $n) $temp1) (fib $temp1 $temp2) (lispval (- $n 2) $temp3)
    (fib $temp3 $temp4) (lispval (+ $temp2 $temp4) $r) (assert ((fib $n $r) /))
  ) )
```

```
YOUR COMMAND =>
((fib 3 $x) space list)
```

```
((fib 3 2) /)
((fib 2 1) /)
((fib 0 0) /)
((fib 1 1) /)
((fib $n $r) (lispval (1- $n) $temp1) (fib $temp1 $temp2) (lispval (- $n 2) $temp3)
(fib $temp3 $temp4) (lispval (+ $temp2 $temp4) $r) (assert ((fib $n $r) /)))
```

```
*** substitutions needed to satisfy your query:
{ $x=2 }
```

I can't find any new justification

```
YOUR COMMAND =>
S
= bye
```

Session 3.10 highlights some of the details involved in using **write**. We should stress that predefined predicates which take arguments, such as **write**, have been defined by **FEXPR**s (using the **DF** defining procedure). Note too that when a variable, say **$x**, is instantiated, character strings which include **$x** are not affected by the instantiation; this is a result of the fact that the definition of the macro-character **$** (see §3.2.4.3) does not apply between two " or | characters: text of type **$x** enclosed within two " or | is retained as it is, thus being unaffected by the instantiations executed by **applysubstit**.

Session 3.10
(microprolog t)

```
YOUR COMMAND =>
((write this and that) (write 'this) (write this | | that))
thisandthat
'this
this that
```

that's correct !

I can't find any new justification

YOUR COMMAND =>
((val (reverse '(a b c)) $y) (write the | | result | | is | | $y))
the result is (c b a)

*** substitutions needed to satisfy your query:
{ $y=(c b a) }

I can't find any new justification

YOUR COMMAND =>
((val (times 5 6 (+ 3 4)) $x) space (write "Here $x : " $x))

Here $x : 210

*** substitutions needed to satisfy your query:
{ $x=210 }

I can't find any new justification

YOUR COMMAND =>
((write $x) (write |$| x) (write "$x"))
$x
$x
$x

that's correct !

I can't find any new justification

YOUR COMMAND =>
((val 100 $x) (write $x) (write |$| x) (write "$x"))
100
$x
$x

*** substitutions needed to satisfy your query:
{ $x=100 }

I can't find any new justification

YOUR COMMAND =>
S
= bye

3.3 A HORN CLAUSE PROVER DIFFERENT FROM PROLOG: REFUTATION BY RESOLUTION, BUT BREADTH FIRST AND POSITIVE UNIT STRATEGY

3.3.1 Return to the strategy employed by PROLOG

In §3.1 we reviewed the principles of theorem proving by refutation, including *refutation by resolution*. §3.2 described in detail the refutation by resolution strategy employed by PROLOG. PROLOG's strategy is defined as a *LUSH resolution* type strategy (hence *linear* and *input*, reserved to sets of *Horn clauses*) but additionally constrained: in particular, the PROLOG inference engine investigates the various resolvents it produces in a *depth first* fashion.

The general strategy of *LUSH resolution* is *refutation complete for binary resolution* (again we must stress **binary** resolution); but it is *nondeterministic*, in the sense that it does not state how some choices are to be operated, especially the choice of what clauses to resolve with each other (and also the choice of literals in these clauses). In contrast, the strategy employed by the PROLOG interpreter is *deterministic*, in that how choices are made is imposed by the language (though some authors speak of nondeterminism in PROLOG because the engine does not automatically pick the best lines at the first attempt but instead has — and is able — to carry out *backtracks*; in our opinion, this amounts to distorting the meaning: see (NILSSON 1971)). By imposing the choice of depth first development (the last resolvent produced is the first to be investigated in the next resolution), PROLOG's (deterministic) strategy loses the completeness of the (nondeterministic) LUSH strategy.

3.3.2   Complete inference rules and complete strategies: breadth first strategy

Earlier on (§3.1.3) we recalled that the *resolution principle* (here: full resolution, not just binary) is a *refutation complete inference rule* in the sense that if a clause set, E (any clauses, not necessarily Horn type), is inconsistent, there exists a finite series of resolutions from these clauses and resolvents which can be derived from them successively which leads to the empty clause.

We must be careful not to confuse a complete inference rule and a complete strategy for implementing that inference rule, even though these notions are related (see (FARRENY & GHALLAB 1986)). The completeness of the resolution principle may be expressed precisely as follows: let $R^1(E)$ or $R(E)$ denote the union of E with the set of all the resolvents obtained by resolving any clauses in E in all possible ways; let $R^0(E) = E$; for $n > 1$ $R^n(E)$ can be defined as the set of resolvents whose parents belong to $R^{n-1}(E)$; the completeness of the resolution principle is equivalent to: $\forall E$ which is an inconsistent set of clauses, $\exists n$ which is a natural integer such that the empty clause belongs to $R^n(E)$. From the hypothesis that E is finite it can be deduced that: $\forall n$, $R^n(E)$ is finite. The set of elements in $R^n(E)$ which do not belong to $R^{n-1}(E)$ can be called: *level n resolvents*.

A strategy imposes choices (either definitive restrictions among the available operations or else an ordering of these operations). Let $R_S^1(E)$ or $R_S(E)$ stand for the union of E with the set of resolvents that can be obtained by resolving any clauses in E in all possible ways but obeying the strategy S; for $n > 1$, $R_S^n(E)$ is defined as the set of resolvents of any clauses in $R_S^{n-1}(E)$. The set of elements in $R_S^n(E)$ which do not belong to $R_S^{n-1}(E)$ will be called: *level n resolvents satisfying strategy S*. The completeness of a strategy S for implementing the resolution principle is equivalent to: $\forall E$

which is an inconsistent set of clauses, $\exists n$ which is a natural integer such that the empty clause belongs to $R_S^n(E)$. A *breadth first strategy* is the name for the general strategy which prevents an element of the resolvents at level $n > 1$ from being produced before all elements at level $n-1$ have been produced. The general breadth first strategy does not discard any resolvent but it does (partially) order their production. From the hypothesis that E is finite, it is simple to deduce that the general breadth first strategy (which, without further specification, is nondeterministic) is complete and that all deterministic implementations of this general strategy are also complete.

3.3.3   Positive unit strategy

The term *unit strategy* is used to describe the general strategy whereby each resolution (not necessarily binary) must involve at least one clause that is reduced to a single literal (called a *unit clause*). The unit strategy is refutation complete for binary resolution (note the word: binary) of Horn clauses, meaning that for every inconsistent set of Horn clauses a refutation exists for which each resolution is binary and involves at least one unit clause. *Positive unit strategy* is the term for the general strategy whereby each resolution must involve a clause reduced to a single *positive* literal (hence an atom, not preceded by the negation connective). The positive unit strategy is also refutation complete for binary resolution of Horn clauses (HENSCHEN 1974) (given this fact, the completeness of simple unit strategy can be immediately deduced).

The (general) unit strategy and the (general) positive unit strategy are nondeterministic in the sense of the previous section since they do not completely impose the choices of clauses and literals for each resolution. In order for these general strategies to be actually implemented, the way in which the choices are to be made must be specified: various individual deterministic unit (or positive unit) strategies can then be defined. All unit strategies necessarily reduce the space of resolvents that can be derived from the initial clause set. If we select unit strategies that use binary resolution alone, each resolvent is guaranteed to contain one literal less than whichever of its parents had the most (a heuristically appealing property since we are approaching a resolvent without any literal!); if in addition the initial clause set only contains Horn clauses, each resolvent will also be a Horn clause.

Fig. 3.6 below shows a *refutation graph* which respects the positive unit strategy (without specifying what substitutions are carried out but with the literals involved underlined). This graph demonstrates the inconsistency of the set of (Horn) clauses:

    { FAIR(mark), DARK(john), FATHER(peter,john),
      FATHER(mark,peter),
      FATHER(john,robert), FATHER(X,Y) $\lor\Box$ SON(Y,X),
      GRANDFATHER(X,Y) $\lor\Box$ FATHER(X,Z) $\lor\Box$ FATHER(Z,Y),
      SON(mark,george),
      $\Box$GRANDFATHER(X,Y) $\lor\Box$ FAIR(X) }

which has previously been studied in §3.2.2. Such a graph can be extracted

from the *search graphs* obtained by applying various (deterministic) particu-
larisations of the general positive unit strategy.

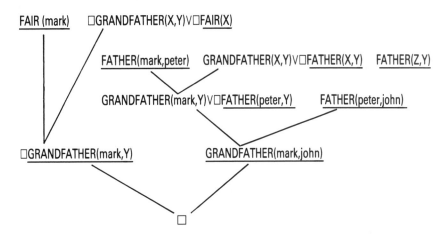

Fig. 3.6.

3.3.4   Breadth first positive unit strategy

Let us take a particular instance of the positive unit strategy (PUS) by stating
that, ∀n, an element at *level n satisfying PUS* cannot be produced until all
those at *level n-1 satisfying PUS* have been produced; this gives us a *breadth
first positive unit strategy*. Taking account of the properties indicated above
pertaining to both breadth first and positive unit strategies, it is clear that the
breadth first positive unit strategy is refutation complete for binary resolution of Horn clauses. In a general fashion, it can be seen that superimposing
the constraint of developing resolvents breadth first onto a complete
strategy results in complete strategy.

As defined below, the breadth first positive unit strategy (BFPUS) is still
nondeterministic since precisely how each successive level is produced is not
stipulated. Later on we propose a program which embodies a particular
implementation, namely a deterministic one. Any individual (deterministic)
implementation of the BFPUS strategy is clearly complete since sooner or
later (but within a finite number of resolutions) it will supply the whole of
level n satisfying S, a level in which the empty clause is bound to appear.

Unlike the PROLOG-type refuters presented in §3.2, the refuter laid
out below will thus be complete for Horn clauses. This theoretical efficiency
may perhaps be limited by the space and time resources required to execute
the program. To improve practical efficiency, the program starting in §3.3.5
could be refined in a variety of ways; we shall make some suggestions at the
end of the description.

3.3.5   Representation of (Horn) clauses

Let E be a set of Horn clauses whose inconsistency we wish to prove. E will be given to the program as a list of clauses, each one being represented by a list of literals; the first element of a clause-list is reserved for the positive literal of the clause; if no such element exists, this element will be (); the following literals, if any, in the clause are negative literals. Finally, each literal is an atom represented by a list consisting of the predicate symbol followed by the arguments, as in §3.1 and 3.2; the atom's place in the clause-list indicates whether the atom is negated or not.

Here, for example, is a clause set E in a form acceptable to the program:

```
( ((SICK a))
  ((LIKES a $X) (DOC $X))
  (() (SICK $X) (QUACK $X) (LIKES $X $Y))
  ((DOC b))
  ((QUACK b)))
```

In the classical formalism (see §3.1), E would be represented by:

$$\{ \text{SICK(a), LIKES(a,X)} \lor \square \, \text{DOC(X),}$$
$$\square \, \text{SICK(X)} \lor \square \, \text{QUACK(Y)} \lor \square \, \text{LIKES(X,Y),}$$
$$\text{DOC(b), QUACK(b)} \}$$

This set of clauses comes from formalising the following problem: 'there exist sick people who like all doctors; no sick person likes quacks; shows that no doctor is a quack'.

Implementation details

To clarify the following text and sessions, we shall agree that when supplying the set E capitals will be used for the symbols of predicates and variables (for example: SICK, $X); but the program given will not worry about maintaining the distinction between upper and lower case letters which may appear in the input symbols (this distinction is not compulsory anyway). Keeping this distinction in Le-Lisp would merely require: (SETQ # :SYSTEM:READ-CASE-FLAG T). But beware: once this system variable has been set to T, subsequent function definitions will have to use lower case for the symbols of primitives used (for fear of not being recognised as defined) until such time as the system variable is reset to its initial value ().

Once the refuter-program is invoked, E's clauses will be split up between 2 identifiers, thenewposu and thenewnonposu; the former will take the list (*one* list, apart from a permutation) of *the new* positive-unit clauses, the latter will take the (one) list of *the* other *new* clauses (*non* positive-unit). To be more exact, thenewposu and thenewnonposu will each take a list of clause *labels* (positive-unit for the first one, non positive-unit for the second). For each clause, an original label of the form g_i (for instance: g101, g102, ...) will be created by calling the GENSYM function; each clause has a label created for it and the list which stands for the clause is assigned to this label. For example, assignment of the value ((SICK a)) to the label g101. This is what the procedure init (initialisation) does:

```
(DE init (e) ;thenewposu, thenewnonposu, theoldposu, theoldnonposu, are
      ;global for init.
    (PROG (name)
        (SETQ theoldposu NIL theoldnonposu NIL thenewposu NIL
        thenewnonposu NIL)
        (MAPC '(LAMBDA (x)
                (COND (  (NULL (CDR x))
                         (SETQ name (GENSYM))
                         (SET name x) (NEWL thenewposu name))
                      (T (SETQ name (GENSYM))
                         (SET name x)
                         (NEWL thenewnonposu name))))
        e)))
```

The part played by theoldposu and theoldnonposu will be explained
later on. Each time init is called they are set to NIL; the same is true for
thenewposu and thenewnonposu.

3.3.6   Creation of any level resolvents

In the description of the procedure newlevel given below, the variables
thenewposu (*the new* positive-unit clauses), theoldposu (*the old* positive-units), thenewnonposu (*the new non* positive-units) and theoldnonposu (*the old non positive* units) are global (they are declared by
hcsrbfpu). Note: before any refutation is invoked, init resets theoldposu,
theoldnonposu, thenewposu and thenewnonposu to NIL (several refutations may be performed in the course of executing hcsrbfuu; the above 4
variables must be reinitialised on each occasion). When newlevel is called
for the nth time, the union of the contents of thenewposu and theoldposu
will hold all the labels of the positive unit clauses of $R^{n-1}_{BFPUS}(E)$, where E
is the initial clause set, while the union of the contents of thenewnonposu
and theoldnonposu will hold all the labels of the non positive unit of
$R^{n-1}_{BFPUS}(E)$. The contents of thenewposu plus those of thenewnonposu will cover all labels at level n-1 (satisfying the breadth first unit strategy
BFPUS). Thus the first time newlevel is called the union of thenewposu
and thenewnonposu will represent exactly E (i.e. $R^0_{BFPUS}(E)$), while
theoldposu and theoldnonposu will be empty. Broadly speaking, newlevel's function is to define and order (via labels) the elements of level n
satisfying BFPUS.

newlevel can be exited on returning 3 types of values. If it is established
(by collect, called from newlevel at either line 4 or 5) that the empty clause
appears in level n, first collect and then newlevel are exited by returning a
label g_i; g_i is then the label for the occurrence of the empty clause obtained
by resolution (and the value of g_i is: (())). The procedure newlevel returns
the symbol failure (line 6) if level n turns out to be empty. Lastly, it returns
the empty list (value NIL) in all other cases (line 9), namely when level n is
not empty but does not contain the empty clause.

The level n resolvents satisfying the breadth first positive unit strategy
(BFPUS) can be obtained by carrying out all possible resolutions between
all the (positive unit) clauses of $R^{n-1}_{BFPUS}(E)$ and all the negative literals of
all clauses in $R^{n-1}_{BFPUS}(E)$, *but only keeping the new resolvents.*

In fact, in order to get level n satisfying BFPUS it is necessary and sufficient to perform resolutions 1) between all the positive unit clauses at level n-1 on the one hand and all the non positive unit clauses of $R^{n-1}{}_{BFPUS}(E)$ on the other hand, and 2) between all the non positive unit clauses at level n-1 on the one hand and all the positive unit clauses of $R^{n-2}{}_{BFPUS}(E)$ on the other hand (note: n-2, so as not to repeat the resolutions between level n clauses). If the restriction defined by points 1 and 2 above is not imposed then the proportion of useless resolutions increases: useless either because they are tried but fail, or else because they succeed but only produce *already existing* resolvents.

However, it is possible for two *distinct* resolutions to lead eventually to the same resolvent; more generally, it is possible for resolvents to be identical in all respects other than their respective variables' names. The program given here does not try to detect these situations. This decision, which reduces the immediate effort involved in execution (comparison of new resolvents with old ones is avoided), could turn out to be costly in the long run: it risks increasing the frequency of useless (i.e. redundant) resolvents.

Line 2 of newlevel declares several variables local to newlevel (but global to collect, a procedure which is called by newlevel and is described later on, and global to order, which is called by collect); these are: resolvent, otheru+ (other positive unit clauses), other≠u+ (*other non positive units*) and allnonposu (*all non positive units*).

Line 3 gathers together all the non positive unit clauses of $R^{n-1}{}_{BFPUS}(E)$ (presuming that the nth call of newlevel is being considered). The call of collect in line 4 sets off all resolutions between all the non positive unit clauses at level n-1 and all the non positive unit clauses of $R^{n-1}{}_{BFPUS}(E)$. Similarly, line 5 sets off all resolutions between all the positive unit clauses of $R^{n-2}{}_{BFPUS}(E)$ and all the non positive unit clauses at level n-1. By calling order, the collect procedure undertakes to store the positive unit and non positive unit clauses at level n in otheru+ and other≠u+ respectively. Line 6 tests whether level n is empty, in which case the refutation fails. In lines 7 and 8 theoldposu, thenewposu, theoldnonposu and thenewnonposu are updated in such a way that when newlevel is exited the union of theoldposu and thenewposu will encompass the positive units of $R^n{}_{BFPUS}(E)$, while the union of theoldnonposu and thenewnonposu will encompass the non positive unit clausess of $R^n{}_{BFPUS}(E)$.

```
1   (DE newlevel () ;Called by invoke
      ;thenewposu, thenewnonposu, theoldposu, theoldnonposu are declared in
      ;hcsrbfpu and initialised in init (to NIL on each refutation)
2     (PROG (resolvent otheru+ other≠u+ allnonposu)
3       (SETQ allnonposu (APPEND thenewnonposu theoldnonposu))
4       (collect thenewposu allnonposu allnonposu)
5       (collect theoldposu thenewnonposu thenewnonposu)
6       (COND ((AND (NULL otheru+) (NULL other≠u+)) (RETURN 'failure)))
```

```
7        (SETQ theoldposu (APPEND thenewposu theoldposu) thenewposu
             otheru+
8              theoldnonposu allnonposu thenewnonposu other≠u+)
9        (RETURN NIL)))
```

The third argument to collect, namely **refnonu** (reference to non unit clauses), serves to reinitialise the value of the≠u+ each time this list has been emptied. Each resolvent is obtained through the procedure **resolution** (the most significant task of this procedure is to compute the resolvent clause — if the unification succeeds — and then to assign it a g_i type label). Once a resolvent has been obtained it is tested for the empty clause. If this is not the case, **order** is asked to memorise it (without testing whether this is a new resolvent).

```
(DE collect (theu+ the≠u+ refnonu) ;Called by newlevel
      ;resolvent is a global variable declared in newlevel
      ;the RETURN relates to the PROG of newlevel
         (WHILE theu+
             (WHILE the≠u+
                 (FOR (i 1 1 (1- (LENGTH (EVAL (CAR the≠u+)))))
                     (SETQ resolvent (resolution
                     (CAR theu+)(CAR the≠u+) i))
                     (COND ((EQUAL '(()) (EVAL resolvent))
                     (RETURN resolvent))
                          (resolvent (order))))
                 (NEXTL the≠u+))
             (SETQ the≠u+ refnonu)
             (NEXTL theu+)))
```

The **order** procedure determines whether or not the given resolvent is a positive unit clause. If so, the resolvent is incorporated into otheru+. If not, the resolvent is incorporated into other≠u+ (non positive unit resolvents).

```
(DE order () ;Called by collect
;resolvent is a global variable declared in newlevel, set in collect
;otheru+, other≠u+ are global variables declared and initialised in newlevel
      (COND ((NULL (CDR (EVAL resolvent))) (NEWL otheru+ resolvent))
            (T (NEWL other≠u+ resolvent))))
```

3.3.7   Representation and evolution of the search graph

Consider the procedure **resolution** given below. Its first argument is a g_i label of a positive unit clause; the second is a g_j label of a non positive unit clause; the third is the index, **k**, of the negative literal (belonging to the non positive unit clause) which is the target of the unification. This k^{th} negative literal is extracted in line 3. At line 4 the result of the unification is stored in the variable **substitutions**. The procedures **unification** and **applysubstit** (as well as the procedures they call) are exactly the same as those given and examined throughout this chapter, starting at §3.1. The **rename** procedure, whose task is to distinguish variables of literals undergoing unification, is no longer the same as the one used in PROLOG interpreter programs (the new version is supplied below). At line 6, an original *label* g_r is created to hold the resolvent's expression.

This expression is formed and assigned to g_r in line 7. In line 8 the triplet (g_i g_j k) is stored as the value of a property, **parents**, attached to g_r; the information may then be used later on to explain the refutation. With the same objective of eventual justification, line 9 saves the list of elementary substitutions required to obtain g_r as the value of a property, **mgu**, attached to g_r.

```
1   (DE resolution (u+ nonu+ n) ;Called by collect
2   (PROG (literal substitutions child)
3      (SETQ literal (NTH n (EVAL nonu+))
4            substitutions (unification (rename (CAR (EVAL u+))) literal))
5      (COND ((EQ 'failure substitutions) (RETURN NIL)))
6      (SETQ child (GENSYM))
7      (SET child (applysubstit substitutions (REMOVE literal (EVAL nonu+))))
8      (PUTPROP child (LIST u+ nonu+ n) 'parents)
9      (PUTPROP child substitutions 'mgu)
10     (RETURN child)))
```

Renaming variables

The refuters described in the first part of this chapter obeyed an *input* strategy (like any PROLOG interpreter), meaning that one of the two clauses to be unified had to be an input clause. As a result, in order to distinguish the variables of two clauses to be unified all that was necessary (see §3.2.3) was to put a number incremented on each cycle after all variable symbols in the input clause (as a *suffix*). For example, the input clause whose (internal) representation was (P ($*$ u) ($*$ v)) was transformed, *at the instant* when the first unification was attempted, into (P ($*$ u 1) ($*$ v 1)), then (P ($*$ u 2) ($*$ v 2)) *at the instant* of the second attempt, etc., always maintaining the same text (P ($*$ u) ($*$ v)) for the input clause to refer back to. The texts of resolvents, unlike those of input clauses, could contain variables with a suffix, but they were not resubmitted to renaming.

This situation has now changed. As a general rule neither of the 2 clauses involved in a unification is an input clause; a priori, both can contain variables with suffixes. For example, we might wish to unify the literals (Q ($*$ u) ($*$ u 4)) and (Q a (f ($*$ u 4))). In the program presented below it has been decided to rename only those variables of the positive unit clause involved in the attempted unification; let (Q ($*$ u) ($*$ u 4)) be this clause. A numeric index (called **index**) is used which is changed on each unification, as in the previous refuter programs (but not simply by incrementing by 1). Assume that **index** equals 8 at the moment when an attempt is made to unify the two literals above; the definition of the value of **index** means it is guaranteed that no variable of the form ($*$ var n), with n $\geq$ 8, yet exists, and in particular there can be no variable of the form $(* u n)$, with $n \geqslant 8$. When the unification takes place it will be dealing with $(Q (* u 8) (* u 12))$ in place of $(Q (* u) (* u 4))$; thus the value of index is added to each suffix (a missing suffix is equivalent to one of zero); for the next unification, index will be set to $12 + 1 = 13$ (12 being the maximum suffix value used). Note that under the old renaming system $(Q (* u) (* u 4))$ would have become $(Q (* u 8) (* u 8))$, which would have introduced confusion between the two variables in the literal. Here is the new code of rename and renam1 (plus that of singu, which is called by renam1; the differences from the version used up until now are underlined):

```
(DE rename (e) ;Called by resolution. index is declared and initialised
;in invoke. the final value of num is set by renam1, via singu
    (PROG (num temp)
        (SETQ num 0 temp (renam1 e) index (+ 1 num index))
        (RETURN temp)))

(DE renam1 (e) ;Called by rename
    (COND ((ATOM e) e)
          ((EQUAL (CAR e) '*) (singu (CDR e)))
          ((MAPCAR 'renam1 e))))

(DE singu (var) ;Called by renam1
;index declared and initialised in invoke, updated in rename
;num declared in rename
    (COND ((NULL (CDR var)) (LIST '* (CAR var) index))
          (T (SETQ num (MAX num (CADR var)))
             (LIST '* (CAR var) (+ index (CADR var))))))
```

3.3.8   First version of the refuter

Program 3.6 below must be invoked by calling hcsrbfpu (**H**orn **c**lause **s**et **r**efuter using a **b**readth **f**irst **p**ositive **u**nit strategy). It incorporates the procedures newlevel, collect, order, resolution, rename, renam1 and singu presented above. The procedure hcsrbfpu is similar to the micropro-log procedures used on several occasions since the beginning of the chapter (from §3.2.3 onwards). The same is true for prompt.

The macro-character $ has also been defined (§3.2.4.3), as before, to make the initial clause set easier to grasp.

Finally, remember that the same unification procedure and the same utility procedures of the various PROLOG interpreters examined in the early part of the chapter are still used here. It is the invoke procedure that, ultimately, orders and controls the production of levels 1, 2, ... n.

Program 3.6

```
(DE hcsrbfpu () ;Main procedure normally called by the user
  (PROG (command set thenewposu thenewnonposu theoldposu
  theoldnonposu)
    (UNTIL NIL
      (prompt "YOUR COMMAND")
      (SETQ command (READ))
      (COND (  (EQ 'S command) (RETURN 'bye))
            (  (EQ '? command)
               (PRINT "The commands understood are:")
               (PRINT "NS,OS (New or Old Set of clauses)")
               (PRINT "SS (Save Set of clauses), E (LISP Evaluation)")
               (PRINT "S (Stop) and ?"))
            (  (EQ 'NS command) (prompt "text of New Set of clauses ?")
               (SETQ set (READ)) (init set) (invoke))
            (  (EQ 'OS command)
               (prompt "name of Old Set of clauses ?")
               (SETQ set (EVAL (READ))) (init set) (invoke))
            (  (EQ 'SS command)
               (prompt "Save Set by what name ?") (SET (READ) set))
            (  (EQ 'E command)
               (prompt "Evaluate what expression ?")
               (PRINT (EVAL (READ))))
            (T (PRINT "Unknown command"))))))

(DMC $ () (LIST '* (READ)))

(DF prompt message ;Called by hcsrbfpu
    (TERPRI) (PRINT (CAR message) " => "))

(DE init (e) ;Initialisation. Called by hcsrbfpu.
  (PROG (name)
    (SETQ theoldposu NIL theoldnonposu NIL thenewposu NIL
          thenewnonposu NIL)
    (MAPC '(LAMBDA (x)
            (COND (  (NULL (CDR x))
                     (SETQ name (GENSYM)) (SET name x)
                     (NEWL thenewposu name))
                  (T (SETQ name (GENSYM))
                     (SET name x) (NEWL thenewnonposu name))))
          e)))

(DE invoke () ;Called by hcsrbfpu
  (PROG (index result)
    (SETQ index 1)
    (UNTIL (SETQ result (newlevel)))
    (COND (  (EQ result 'failure)
          (  (TERPRI)
          (  (PRINT "I can't find any justification"))
          (T (PRINT "Refutation succeeded " result)))))
```

```
(DE newlevel () ;Called by invoke
;thenewposu, thenewnonposu, theoldposu, theoldnonposu are declared in
;hcsrbfpu and initialised in init (to NIL on each refutation)
   (PROG (resolvent otheru+ other≠u+ allnonposu)
      (SETQ allnonposu (APPEND thenewnonposu theoldnonposu))
      (collect thenewposu allnonposu allnonposu)
      (collect theoldposu thenewnonposu thenewnonposu)
      (COND ((AND (NULL otheru+) (NULL other≠u+))
            (RETURN 'failure)))
      (SETQ theoldposu (APPEND thenewposu theoldposu)
            thenewposu otheru+
            theoldnonposu allnonposu thenewnonposu other≠u+)
      (RETURN NIL)))

(DE collect (theu+ the≠u+ refnonu) ;Called by newlevel
;resolvent is a global variable declared in newlevel
;the RETURN relates to the PROG of newlevel
   (WHILE theu+
      (WHILE the≠u+
         (FOR (i 1 1 (1- (LENGTH (EVAL (CAR the≠u+)))))
               (SETQ resolvent (resolution (CAR theu+) (CAR the≠u+) i))
               (COND ((EQUAL '(()) (EVAL resolvent)) (RETURN resolvent))
                     (resolvent (order))))
         (NEXTL the≠u+))
      (SETQ the≠u+ refnonu)
      (NEXTL theu+)))

(DE order () ;Called by collect
;resolvent is a global variable declared in newlevel, set in collect
;otheru+, other≠u+ are global variables declared and initialised in newlevel
   (COND ((NULL (CDR (EVAL resolvent))) (NEWL otheru+ resolvent))
         (T (NEWL other≠u+ resolvent))))

(DE resolution (u+ nonu+ n) ;Called by collect
   (PROG (literal substitutions child)
         (SETQ literal (NTH n (EVAL nonu+))
               substitutions (unification (rename (CAR (EVAL u+))) literal))
         (COND ((EQ 'failure substitutions) (RETURN NIL)))
         (SETQ child (GENSYM))
         (SET child (applysubstit substitutions (REMOVE literal (EVAL
               nonu+))))
         (PUTPROP child (LIST u+ nonu+ n) 'parents)
         (PUTPROP child substitutions 'mgu)
         (RETURN child)))
```

```
(DE rename (e) ;Called by resolution. index is declared and initialised
;in invoke. the final value of num is set by renam1, via singu
      (PROG (num temp)
            (SETQ num 0 temp (renam1 e) index (+ 1 num index))
            (RETURN temp)))

(DE renam1 (e) ;Called by rename
      (COND ((ATOM e) e)
            ((EQUAL (CAR e) '*) (singu (CDR e)))
            ((MAPCAR 'renam1 e))))

(DE singu (var) ;Called by renam1
;index declared and initialised in invoke, updated in rename
;num declared in rename
      (COND (   (NULL (CDR var)) (LIST '* (CAR var) index))
            (T (SETQ num (MAX num (CADR var)))
               (LIST '* (CAR var) (+ index (CADR var))))))

(DE unification (exp1 exp2) ;Called by resolution
      (PROG (d d1 d2 mmgu)
            (COND ((NOT (EQUAL (len exp1) (len exp2))) (RETURN 'failure)))
            (UNTIL (NULL (SETQ d (disagreement exp1 exp2)))
                  (COND (   (variable? d1)
                            (COND ((in? d1 d2) (RETURN 'failure))
                                  ((SETQ mmgu (CONS d mmgu)
                                        exp1 (SUBST d2 d1 exp1)
                                        exp2 (SUBST d2 d1 exp2)))))
                        (   (variable? d2)
                            (COND ((in? d2 d1) (RETURN 'failure))
                                  ((SETQ mmgu (CONS (REVERSE d) mmgu)
                                        exp1 (SUBST d1 d2 exp1)
                                        exp2 (SUBST d1 d2 exp2)))))
                        (T (RETURN 'failure))))
            (REVERSE mmgu)))

(DE applysubstit (listsubs listexpr) ;Called by resolution
      (UNTIL (NULL listsubs)
            (SETQ listexpr (SUBST (CADAR listsubs) (CAAR listsubs) listexpr))
            (SETQ listsubs (CDR listsubs)))
      listexpr)

(DE len (x) ;Called by unification
      (COND ((ATOM x) 1)
            ((EQUAL (CAR x) '*) 1)
            ((LENGTH x))))
```

```
(DE disagreement (exp1 exp2) ;Called by unification and disagreement
        (COND ( (EQUAL exp1 exp2) NIL)
                ( (OR (ATOM exp1) (ATOM exp2) (EQ (CAR exp1) '*)
                                               (EQ (CAR exp2) '*))
                  (LIST exp1 exp2))
                ( (NOT (EQUAL (CAR exp1) (CAR exp2)))
                  (disagreement (CAR xp1) (CAR exp2)))
                ( (disagreement (CDR exp1) (CDR exp2)))))

(DE variable? (I) ;Called by unification
        (COND ((ATOM I) ()) ((EQUAL (CAR I) '*))))

(DE in? (I1 I2) ;Called by unification and in?
        (COND ((ATOM I2) NIL)
                ((EQUAL I1 I2))
                ((OR (in? I1 (CAR I2)) (in? I1 (CDR I2))))))
```

Session 3.11 is obtained with program 3.6 already loaded. The clause set
supplied to the refuter is that examined in §3.3.5 (sick people, doctors and
quacks). The values associated with resolvents' labels and with the **parents**
and **mgu** properties can be consulted both from within **hcsrbfpu** and after
calling it: this allows the detail of the refutation to be followed; we have
given an outline of what can be done below. Program 3.6 will be completed
later on (starting at program 3.7, first see §3.3.9) so that it will be able to
explain the refutations found.

Session 3.11
(hcsrbfpu)

YOUR COMMAND =>
?
The commands understood are:
NS,OS (New or Old Set of clauses)
SS (Save Set of clauses), E (LISP Evaluation)
S (Stop) and ?

YOUR COMMAND =>
NS

text of New Set of clauses ? =>
(((SICK a)) ((LIKES a $X) (DOC $X))
  (() (SICK $X) (QUACK $Y) (LIKES $X $Y)) ((DOC b)) ((QUACK b)))
Refutation succeeded g117

YOUR COMMAND =>
E

```
Evaluate what expression ? =>
g117
(())

YOUR COMMAND =>
E

Evaluate what expression ? =>
(GET 'g117 'parents)
(g110  g116  1)

YOUR COMMAND =>
S
= bye

g110
= ((likes a b))
g116
= (() (likes a b))
(GET 'g110 'parents)
= (g107  g105  1)
g107
= ((doc b))
g105
= ((likes a (* x)) (doc (* x)))
(GET 'g111 'mgu)
= (((* x) b))
```

The series of resolutions can be broken down bit by bit into the precise steps which led to the empty clause (g117), thereby reconstituting the refutation graph. More generally, the search graph (which includes all the resolvents produced and not just those involved in the refutation graph) can be reconstructed by examining in turn every g_i for $i = p+1$ to 117, where p represents the original value of the counter associated with the GENSYM primitive; for this session, $p = 103$ (a particular value of the Le-Lisp interpreter used).

3.3.9   Procedure for reconstituting and displaying the refutation

To make it easier to understand the improvements to program 3.1 that are to be integrated into the next version of the refuter, let us consider in isolation the problem of how to reconstitute and display the refutation (potentially) executed when hcsrbfpu is called. A refutation graph is represented symbolically in Fig. 3.7 below: e, f, h, i, j and d are input clause labels; k, a, b, x, y and gNIL are resolvent labels; gNIL is the label of the empty clause.

Let us assume that this refutation graph was achieved through LISP atoms e, f, ... gNIL for the nodes, and through the property value parents associated with each atom e, f, ... gNIL for the arcs. Hence to obtain the graph we could do the following:

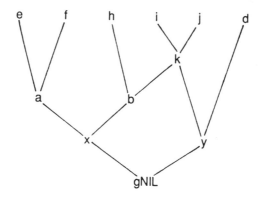

Fig. 3.7.

```
(PUTPROP 'gNIL '(x y) 'parents)
(x y)
(PUTPROP 'x '(a b) 'parents)
= (a b)
(PUTPROP 'y '(k d) 'parents)
= (k d)
(PUTPROP 'a '(e f) 'parents)
= (e f)
(PUTPROP 'b '(k h) 'parents)
= (k h)
(PUTPROP 'k '(i j) 'parents)
= (i j)
```

For example, the parents of gNIL are x and y. This representation is close to that produced by program 3.6: at the end of a successful execution of the program we are left with a label like gNIL whose parents and ancestors can be discovered via the successive values of the **parents** property for gNIL, then for its parents which have just been obtained, etc. ... In order to be able to explain how gNIL has been reached, let us first set ourselves the goal of constructing the list:

$$(gnil (x (a e f) (b (k i j) h)) (y (k i j) d))$$

which is a precise and straightforward representation of the refutation graph in Fig. 3.4. Procedure **tree** below performs the desired function given a single argument: the starting atom (clause label for our refuter-program).

```
(DE tree (lab)
    (PROG (temp)
       (SETQ temp (GET lab 'parents))
       (COND ((NULL temp) lab)
             (T (LIST lab (tree (CAR temp)) (tree (CADR temp)))))))
```

You can check that:

```
(tree 'gnil)
= (gnil (x (a e f) (b (k i j) h)) (y (k i j) d))
```

But this kind of coding leads to redundancy: for instance, in the above expression it can be seen that b's parents are k, whose parents are i and j, and h, which is an input atom; it can also be seen that y's parents are k, whose parents are i and j, and d, which is an input atom. It is pointless, and indeed tiresome, to supply the ancestry of an atom (here k) all over again if it has already be given once. For example, it would be better to show the preceding refutation graph as:

(gnil (x (a e f) (b (k i j) h)) (y k d))

The second version of tree avoids the redundancies permitted in the first by using the extra list already to store labels whose parents have previously been displayed.

```
(DE tree (lab) (PROG (already) (tree1 lab)))
(DE tree1 (lab) ;Called by tree and tree1
    (PROG (temp)
        (COND (  (MEMBER lab already) lab)
              (T (SETQ temp (GET lab 'parents))
                 (COND ((NULL temp) lab)
                       (T (NEWL already lab)
                          (LIST lab (tree1 (CAR temp))
                                    (tree1 (CADR temp)))))))))
```

It can be seen that:
```
(tree 'gnil)
= (gnil (x (a e f) (b (k i j) h)) (y k d))
```

To explain how we arrive at, say, gnil from its ancestors, we have only to enter (tell (tree 'gNIL)) where tell is defined as follows:

```
(DE tell (x)
    (COND (  (AND (ATOMP (CADR x)) (ATOMP (CADDR x)))
             (PRINT (CAR x) " obtained by resolving "
                    (CADR x) " and " (CADDR x)))
          (  (ATOMP (CADR x))
             (tell (CADDR x))
          (  (PRINT (CAR x) " obtained by resolving "
          (        (CADR x) " and " (CAADDR x)))
          (  (ATOMP (CADDR x))
          (  (tell (CADR x))
          (  (PRINT (CAR x) " obtained by resolving "
          (        (CAADR x) " and " (CADDR x)))
          (T (tell (CADR x))
          (  (tell (CADDR x))
          (  (PRINT (CAR x) " obtained by resolving "
          (        (CAADR x) " and " (CAADDR x)))))))
```

Here is the result of calling tell:

```
(tell '(gnil (x (a e f) (b (k i j) h)) (y k d)))
a obtained by resolving e and f
k obtained by resolving i and j
b obtained by resolving k and h
x obtained by resolving a and b
y obtained by resolving k and d
gnil obtained by resolving x and y
= y
```

Now let us modify the example. The value of an atom's **parents** property will be NIL if that atom has no parents: in this case let us agree that the atom represents an input clause. Otherwise, the value of this property will be a triplet formed from two parent-atoms and one number: in this case let us agree that the first parent-atom represents a positive unit clause, the second parent-atom represents a non positive unit clause and the number represents the index in the clause of the negative literal unified with the first clause. Program 3.6 already allowed for the value of the **parents** property of each resolvent clause to be supplied in this way. In the following example the clause symbolised by gNIL represents the resolvent of clause x with the second negative literal of the clause symbolised by y.

```
(PUTPROP 'gNIL '(x y 2) 'parents)
= (x y 2)
(PUTPROP 'x '(a b 1) 'parents)
= (a b 1)
(PUTPROP 'y '(k d 3) 'parents)
= (k d 3)
(PUTPROP 'a '(e f 2) 'parents)
= (e f 2)
(PUTPROP 'b '(k h 4) 'parents)
= (k h 4)
(PUTPROP 'k '(i j 2) 'parents)
= (i j 2)
```

To make use of the new format of the values taken by parents, slight changes to tree1 and tell are required; at the same time, we have condensed the code of tell. The alterations are underlined.

```
(DE tree1 (lab) ;Called by tree and tree1 (PROG (temp)
    (COND (   (MEMBER lab already) lab)
            (T (SETQ temp (GET lab 'parents))
               (COND ((NULL temp) lab)
                    (T (NEWL already lab)
                       (LIST lab
                            (tree1 (CAR temp))
                            (tree1 (CADR temp))
                            (CADDR temp)))))))))
```

We can immediately confirm that:

```
(tree 'gnil)
= (gnil (x (a e f 2) (b (k i j 2) h 4) 1) (y k d 3) 2)
```

```
(DE tell (x)
    (COND (   (AND (ATOMP (CADR x)) (ATOMP (CADDR x)))
              (write x (CADR x) (CADDR x)))
          (   (ATOMP (CADR x)) (tell (CADDR x))
              (write x (CADR x) (CAADDR x)))
          (   (ATOMP (CADDR x)) (tell (CADR x))
              (write x (CAADR x) (CADDR x)))
          (T (tell (CADR x))
             (tell (CADDR x))
             (write x (CAADR x) (CAADDR x))))))
(DE write (x y z) ;Called by tell
    (PRINT (CAR x) " obtained by resolving " y
            " and literal " (CADDDR x) " of " z))
```

As an example, here is the new effect of calling (tell (tree 'gnil)):

```
(tell (tree 'gnil))
a obtained by resolving e and literal 2 of f
k obtained by resolving e and literal 2 of j
b obtained by resolving e and literal 4 of h
x obtained by resolving e and literal 1 of b
y obtained by resolving e and literal 3 of d
gnil obtained by resolving e and literal 2 of y
= y
```

3.3.10   Second version of the refuter (including display of the refutation)

In order to be able to explain which resolutions lead to the empty clause being derived, a new version (program 3.7) of refuter-program 3.6 is now defined. This version includes procedures similar to the last versions of tree, tree1 and tell presented above. Before going into the details of the modifications to the code that change program 3.6 into program 3.7, we first show a working session (session 3.12) which is obtained by loading program 3.7 and giving the variable sickpeople the value:

```
( ((SICK a)) ((LIKES a $X) (DOC $X))
  (() (SICK $X) (QUACK $Y) (LIKES $X $Y)) ((DOC b)) ((QUACK b)) )
```

(This is still the same set of clauses as those expressing the sick people, doctors and quacks problem of § 3.3.5).

Session 3.12

```
(hcsrbfpu)
YOUR COMMAND =>
OS
name of Old Set of clauses ? =>
sickpeople
Refutation succeeded

g7 = ((likes a b)) obtained by resolving
g4 =((doc b)) and literal 1 of g2 = ((likes a $x) (doc $x))
substituting b for $x;

g6 = (() (sick $x) (likes $x b)) obtained by resolving
g5 = ((quack b)) and literal 2 of g3 = (() (sick $x) (quack $y) (likes $x $y))
substituting b for $y;
```

g13 = (() (likes a b)) obtained by resolving
g1 = ((sick a)) and literal 1 of g6 = (() (sick $x) (likes $x b))
substituting a for $x;

g14 = (()) obtained by resolving
g7 = ((likes a b)) and literal 1 of g13 = (() (likes a b))

YOUR COMMAND =
S
= bye

When the refutation succeeds, program 3.7 displays (starting from the last resolution — the one which led to the empty clause) the detail of those resolutions *that are indispensable* to the refutation; for each one it supplies: (i) the labels and expressions of the parent clauses, (ii) the index of the negative literal in the non unit clause involved in the refutation, and (iii) the substitutions performed. Furthermore, the variables present in the clauses and substitutions are shown in *external form* by means of procedure **paste**, previously defined in § 3.2.4.3. Finally, a slight modification is made to line 3 of init to reinitialise the counter associated with the LISP function **GENSYM** each time the refutation is invoked (an explicit assignment of the system variable that acts as the counter is made; other dialects possess a specific function such as **SETGEN** to do this). Thus each time the refutation is executed at least some of the g_i symbols of earlier calls in the same session will be reused; as a result, every time a label g_i is introduced or reintroduced, the values of the **parents** and **mgu** properties are first wiped out (hence the definition of **wipe**, called in lines 7 and 9 of init).

Notice that the label of the last resolvent (the empty clause) obtained in the previous example is **g14**; **g1** to **g5** are the labels standing for the input clauses. Thus 9 resolvents were produced. The justification shown by the program takes 4 resolutions to reach the empty clause (hence 5 wasted resolutions).

Program 3.7
Compared with program 3.6, the underlined sections mark (i) the new parts of previously written procedures (init, invoke), and (ii) the names of brand new procedures (wipe, tell, verse, writeresolvent, writeparents, writemgu, tree, tree1, paste) Procedures hcsrbfpu, $, prompt, newlevel, collect, order, resolution, rename, renam1, singu, applysubstit, unification, len, disagreement, variable? and in? of program 3.6 have been preserved in program 3.7.

```
(DE init (e) ;Called by hcsrbfpu.
    (PROG (name)
        (SETQ #:SYSTEM:GENSYM-COUNTER 0
                theoldposu NIL theoldnonposu NIL thenewposu NIL
                thenewnonposu NIL)
        (MAPC '(LAMBDA (x)
                (COND (  (NULL (CDR x))
                        (SETQ name (GENSYM))
```

```
            (SET name x) (wipe name) (NEWL thenewposu name))
        (T (SETQ name (GENSYM))
            (SET name x) (wipe name) (NEWL thenewnonposu name))))
                e)))

(DE wipe (n) ;Called by init
    (REMPROP n 'parents) (REMPROP n 'mgu))

(DE invoke () ;Called by hcsrbfpu
    (PROG (index result)
        (SETQ index 1)
        (UNTIL (SETQ result (newlevel)))
        (COND (  (EQ result 'failure)
                (TERPRI)
                (PRINT "I can't find any justification"))
            (T (PRINT "Refutation succeeded") (tell (tree result)))))))

(DE tell (x) ;Called by invoke and tell
    (PROG (p1 p2 lit) ;p1 and p2 for the parent clauses,
                    ;lit for the index of the literal
        (SETQ p1 (CADR x) p2 (CADDR x) lit (CADDDR x))
        (COND (  (AND (ATOMP p1) (ATOMP p2))
                (verse x p1 p2 lit))
            (  (ATOMP p1)
                (tell p2)
                (verse x p1 (CAR p2) lit))
            (  (ATOMP p2)
                (tell p1)
                (  (verse x (CAR p1) p2 lit))
            (T (tell p1)
                (tell p2)
                (verse x (CAR p1) (CAR p2) lit)))))

(DE verse (x p1 p2 lit) ;Called by tell
    (writeresolvent x) (writeparents p1 p2 lit) (writemgu x))

(DE writeresolvent (x) ;Called by verse
    (TERPRI)
    (PRINT (CAR x) " = " (paste (EVAL (CAR x))) " obtained by resolving " ))

(DE writeparents (p1 p2 lit) ;Called by verse
    (PRINT p1 " = " (paste (EVAL p1)) " and literal "
        lit " of " p2 " = " (paste (EVAL p2))))

(DE writemgu (x) ;Called by verse
    (SETQ x (GET (CAR x) 'mgu))
    (COND ( (NOT (NULL x))
            (PRIN " substituting " )
            (MAPC '(LAMBDA (y)
                    (PRIN (paste (CADR y)) " for " (paste (CAR y)) " ; "))
                x)
            (TERPRI))))

(DE tree (lab) ;Called by invoke. See § 3.3.9
    (PROG (already) (tree1 lab)))
```

```
(DE tree1 (lab) ;Called by tree and tree1
    (PROG (temp)
        (COND (   (MEMBER lab already) lab)
                (T (SETQ temp (GET lab 'parents))
                    (COND ((NULL temp) lab)
                        (T (NEWL already lab)
                            (LIST lab
                                (tree1 (CAR temp))
                                (tree1 (CADR temp))
                                (CADDR temp))))))))

(DE paste (e) ;Called by writeresolvent, writeparents, writemgu.
            ;See § 3.2.4.3
    (COND ((ATOMP e) e)
        ((EQ '* (CAR e)) (APPLY 'CONCAT (CONS '|$| (CDR e))))
        (T (MAPCAR (LAMBDA (x) (paste x)) e))))
```

The definition of program 3.8 below includes a fresh enhancement aimed at presenting the justifications for the refutations discovered in a more readable fashion. To make the proposed modifications easier to follow, a session (session 3.13) is presented first; program 3.8 has been loaded before, and the variable sickpeople has been set to the list of clauses:

(((SICK a)) ((LIKES a $X) (DOC $X))
  ((() (SICK $X) (QUACK $Y) (LIKES $X $Y)) ((DOC b)) ((QUACK b)))

Session 3.13

(hcsrbfpu)

YOUR COMMAND =>
OS

name of Old Set of clauses ? =
sickpeople
Refutation succeeded

g7 = (likes a b) obtained by resolving
    g4 = (doc b) and literal 1 of
    g2 = (likes a $x)or-(doc $x)
    substituting b for $x;

g6 = -(sick $x)or-(likes $x b) obtained by resolving
    g5 = (quack b) and literal 2 of
    g3 = -(sick $x)or-(quack $y)or-(likes $x $y)
    substituting b for $y;

g13 = -(likes a b) obtained by resolving
    g1 = (sick a) and literal 1 of
    g6 = -(sick $x)or-(likes $x b)
    substituting a for $x;

g14 = () obtained by resolving
    g7 = (likes a b) and literal 1 of
    g13 = -(likes a b)

YOUR COMMAND =
S
= bye

The clauses are written out in a more readable style: negative literals have an explicit leading — and the literals of a single clause are concatenated with the **or** connective. Remember that "literal i" refers to the *i*th *negative* literal. The reader should find it easy to alter program 3.8 if he wishes to display predicates and terms in the classical form: P(...) or t(...).

Program 3.8 is now given, only modified procedures being shown: all the rest of the procedures are exactly the same as those of program 3.7. Altered sections and new procedure names (relative to program 3.7) are underlined.

Program 3.8

```
(DE writeresolvent (x) ;Called by verse
     (TERPRI)
     (PRIN (CAR x) " = ")
     (pclear (paste (EVAL (CAR x))))
     (PRINT " obtained by resolving "))

(DE pclear (e) ;Called by writeresolvent, writeparents. (print clearly)
     (COND ((NULL (CDR e)) (PRIN (CAR e)))
           (T (COND ((NOT (NULL (CAR e))) (PRIN (CAR e) " or ")))
           (pclear- (CDR e)))))

(DE pclear- (e) ;Called by pclear and pclear- (print negative literals clearly)
     (COND ((NULL (CDR e)) (PRIN '- (CAR e)))
           (T (PRIN '- (CAR e) " or ") (pclear- (CDR e)))))

(DE writeparents (p1 p2 lit) ;Called by verse
     (PRIN " " p1 " = ")
     (pclear (paste (EVAL p1)))
     (PRINT " and literal " lit " of ")
     (PRIN " " p2 " = ")
     (pclear (paste (EVAL p2)))
     (TERPRI))

(DE writemgu (x) ;Called by verse
     (SETQ x (GET (CAR x) 'mgu))
     (COND ( (NOT (NULL x))
           (PRIN " substituting ")
           (MAPC '(LAMBDA (y)
                   (PRIN (paste (CADR y)) " to " (paste (CAR y)) " ; "))
                 x)
           (TERPRI))))
```

A new test for program 3.8

Let us take the following problem: given a set with an associative binary operation £ which always takes solutions on the left and on the right, prove that an identity element exists on the right-hand side for £.

First let us adopt a representation. We shall agree that the predicate

P(x,y,z) takes the value TRUE if and only if x£y=z. The assertions in the problem can be expressed by the set of formulae:

>   ∀x,y ∃z P(z,x,y)   since solutions always exist on the left,

>   ∀x,y ∃z P(x,z,y)   since solutions always exist on the right,

>   ∀x,y,z,u,v,w P(x,y,u) P(y,z,v) P(x,v,w) --> P(u,z,w)   and

>   ∀x,y,z,u,v,w P(x,y,u) P(y,z,v) P(u,z,w) --> P(x,v,w)
                                  since the operation is associative.

We have to prove that: ∃e such that ∀x P(e,x,e); this is equivalent to refuting: ∀e ∃x □P(e,x,e) due to the 4 preceding axioms. Hence we are trying to prove the inconsistency of the set of 5 clauses:

{ P(g(x,y),x,y), P(x,h(x,y),y), P(k(x),x,k(x)),
                □P(x,y,u) ∨ □P(y,z,v) ∨ □P(x,v,w) ∨ P(u,z,w)
                □P(x,y,u) ∨ □P(y,z,v) ∨ □P(u,z,w) ∨ P(x,v,w) }

To attempt to find a refutation with program 3.8, this set will be coded by the following list:

```
( ((P (g $X $Y) $X $Y))
  ((P $X (h $X $Y) $Y))
  ((() (P (k $X) $X (k $X)))
  ((P $U $Z $W) (P $X $Y $U) (P $Y $Z $V) (P $X $V $W))
  ((P $X $V $W) (P $X $Y $U) (P $Y $Z $V) (P $U $Z $W)))
```

Assume that the identifier identright1 has been set to this list and invoke program 3.8.

Session 3.14

(hcsrbfpu)

YOUR COMMAND =>
OS

name of Old Set of clauses ? =>
identright1
Refutation succeeded

g17 = (p $u $z $w)or-(p (g $v $w) $y $u)or-(p $y $z $v)
obtained by resolving
    g1 = (p (g $x $y) $x $y) and literal 3 of
    g4 = (p $u $z $w)or-(p $x $y $u)or-(p $y $z $v)or-(p $x $v $w)
    substituting (g $x13 $y13) for $x; $v for $x13; $w for $y13;

g19 = (p $u (h $y $v) $w)or-(p (g $v $w) $y $u)
obtained by resolving
    g2 = (p $x (h $x $y) $y) and literal 2 of
    g17 = (p $u $z $w)or-(p (g $v $w) $y $u)or-(p $y $z $v)
    substituting $y for $x16; (h $y $y16) for $z; $v for $y16;

g109 = (p $u (h $y $y) $u) obtained by resolving
    g1 = (p (g $x $y) $x $y) and literal 1 of
    g19 = (p $u (h $y $v) $w)or-(p (g $v $w) $y $u)

substituting $v for $x149; $w for $y149; $y for $v; $u for $w;

g170 = () obtained by resolving

    g109 = (p $u (h $y $y) $u) and literal 1 of

    g3 = -(p (k $x) $x (k $x))

    substituting (k $x) for $u225; (h $y225 $y225) for $x;

YOUR COMMAND =>

S

= bye

Each 'verse' details the precise conditions of how a resolution is carried out. The two parent clauses appear as they were *before* the renaming prior to unification. On the other hand, of course, the text of the substitutions does take the renaming of variables into account (recall that renaming is only performed on the positive unit clause).

For example, the first 'verse' states that the variable $x of g1 has been renamed $x13 (so there can be no confusion with the $x variable of g4), as witnessed by the list of substitutions. To help in examining the results, it would be better to display the positive unit clause as it was *after* renaming had taken place. To this end, program 3.9 is derived from program 3.8 with modifications (underlined sections) to procedures resolution, tell, verse, writeparents and tree1, and the addition of procedure ren.

Program 3.9

```
(DE resolution (u+ nonu+ n) ;Called·by collect
  (PROG (literal substitutions child ind)
    (SETQ literal (NTH n (EVAL nonu+))
          ind index
          substitutions (unification (rename (CAR (EVAL ))) literal))
    (COND ((EQ 'failure substitutions) (RETURN NIL)))
    (SETQ child (GENSYM))
    (SET child (applysubstit substitutions (REMOVE literal (EVAL nonu+))))
    (PUTPROP child (LIST u+ nonu+ n ind) 'parents)
    (PUTPROP child substitutions 'mgu)
    (RETURN child)))

(DE tell (x) ;Called by invoke and tell
  (PROG (p1 p2 lit ind)
    (SETQ p1 (CADR x) p2 (CADDR x) lit (CADDDR x) ind (CAR (CDDDDR x)))
    (COND ((AND (ATOMP p1) (ATOMP p2)) (verse x p1 p2 lit ind))
          ((ATOMP p1) (tell p2) (verse x p1 (CAR p2) lit ind))
          ((ATOMP p2) (tell p1) (verse x (CAR p1) p2 lit ind))
          (T (tell p1) (tell p2) (verse x (CAR p1) (CAR p2) lit ind)))))

(DE verse (x p1 p2 lit ind) ;Called by tell
  (writeresolvent x) (writeparents p1 p2 lit ind) (writemgu x))
```

```
(DE writeparents (p1 p2 lit ind) ;Called by verse
    (PRIN " " p1 " = ")
    (pclear (paste (ren (EVAL p1) ind)))
    (PRINT " and literal " lit " of ")
    (PRIN " " p2 " = ")
    (pclear (paste (EVAL p2)))
    (TERPRI))

(DE ren (e i) ;Called by writeparents
    (COND ((ATOM e) e)
          ((EQUAL (CAR e) '*) (COND ((NULL (CDDR e)) (LIST '* (CADR e) i))
                                    (T (LIST '* (CADR e)
                                             (+ i (CADDR e))))))
          ((MAPCAR '(LAMBDA (x) (ren x i)) e))))

(DE tree1 (lab) ;Called by tree
   (PROG (temp)
       (COND (  (MEMBER lab already) lab)
             (T (SETQ temp (GET lab 'parents))
                (COND ((NULL temp) lab)
                      (T (NEWL already lab)
                         (LIST lab
                               (tree1 (CAR temp))
                               (tree1 (CADR temp))
                               (CADDR temp)
                               (CADDDR temp)))))))))
```

The principle behind the change is as follows. In resolution, when a resolvent is created, the value of index which applied at the time renaming took place is stored (in 4th position) in the value of the parents property associated with the label representing the resolvent. At the instant when the program has to write out expression e of a resolvent's first parent (a positive unit clause), the ren procedure is applied to e (the first argument to ren): the index value given as the second argument to ren is appended to the suffix of each variable appearing in e (this action is justified by the way in which renaming is performed: see § 3.3.7).

Session 3.15 below, which is obtained by executing program 3.9, should be compared line by line with session 3.14, which was obtained by executing program 3.8.

Session 3.15

(hcsrbfpu)

YOUR COMMAND =>
OS

name of Old Set of clauses ? =>
identright1
Refutation succeeded

g17 = (p $u $z $w)or-(p (g $v $w) $y $u)or-(p $y $z Sv)
obtained by resolving
    g1 = (p (g $x13 $y13) $x13 $y13) and literal 3 of
    g4 = (p $u $z $w)or-(p $x $y $u)or-(p $y $z $v)or-(p $x $v $w)
    substituting (g $x13 $y13) for $x; $v for $x13; $w for $y13;

g19 = (p $u (h $y $v) $w)or-(p (g $v $w) $y $u)
obtained by resolving
    g2 = (p $x16 (h $x16 $y16) $y16) and literal 2 of
    g17 = (p $u $z $w)or-(p (g $v $w) $y $u)or-(p $y $z $v)
    substituting $y for $x16; (h $y $y16) for $z; $v for $y16;

g109 = (p $u (h $y $y) $u) obtained by resolving
    g1 = (p (g $x149 $y149) $x149 $y149) and literal 1 of
    g19 = (p $u (h $y $v) $w)or-(p (g $v $w) $y $u)
    substituting $v for $x149; $w for $y149; $y for $v; $u for $w;

g170 = () obtained by resolving
    g109 = (p $u225 (h $y225 $y225) $u225) and literal 1 of
    g3 = -(p (k $x) $x (k $x))
    substituting (k $x) for $u225; (h $y225 $y225) for $x;

YOUR COMMAND =
S
= bye

3.3.11   Example applications

All the following examples are dealt with by calling program 3.9. To make them easier to follow and understand, it is assumed that before **hcsrbfpu** is called an identifier (such as **sickpeople** or **identright1** above) is set to the list representing the clause set whose inconsistency we are trying to prove. Naturally, this kind of assignment is not compulsory (the procedure **hcsrbfpu** allows interaction to achieve this).

The clauses are written directly, obeying the format agreed for using the refuter-program (which has been followed in the preceding examples and was defined in § 3.3.5). From now on, the identifiers for the clause sets are defined first, at the same time pointing out the possible problems presented; then all the refutations will be attempted via a single call of **hcsrbfpu**.

(1) Show the inconsistency of the clause set **set1** defined by:

```
(SETQ set1 '( (() (P $X) (Q $X))
              ((P b) (R a $Y))
              ((R a b))
              ((Q a))                ))
```

In fact, this set is consistent. For instance, all interpretations in which (P $x) is always false, other than x=b, while (Q a) and (R a b) are true, satisfy all 4 clauses at the same time.

(2) Show the inconsistency of the clause set **set2** defined by:

```
(SETQ set2 '( (() (P $X) (Q (f $X)))
              ((P b) (R a $Y))
              (R a b))
              ((Q $Z))                ))
```

Although syntactically similar to the previous one, this clause set is actually inconsistent.

(3) Show the inconsistency of the clause set **set3** defined by:

```
(SETQ set3 '( ((P a b))
             ((Q c b))
             ((P d c))
             ((Q e c))
             ((R $X $Y) (P $X $Y))
             ((R $X $Y) (Q $X $Y))
             ((S $X $Z) (P $X $Y) (R $Y $Z))
             ((T $X $Z) (Q $X $Y) (R $Y $Z))
             (() (T e b))                    ))
```

This set is in fact inconsistent.

(4) Show the inconsistency of the clause set **logicians** defined by:

```
(SETQ logicians '( (() (SON $X) (LOGICIAN $X))
                   ((LOGICIAN $X) (SOUND $X))
                   ((SOUND $X) (JUDGE $X))
                   ((SON a))
                   ((JUDGE a))                ))
```

The predicate (**SON** $x) stands for: '$x is one of your sons'; the predicate (**LOGICIAN** $x) stands for: '$x can be a logician'; the predicate (**SOUND** $x) stands for: '$x is an individual who is sound of mind'; the predicate (**JUDGE** $x) stands for: '$x is an individual capable of being a judge'. The first 3 clauses state that: (i) none of your sons can be a logician, (ii) any individual who is sound of mind can be a logician, (iii) all people capable of being judges are sound of mind. The last 2 clauses arise from negating the conjecture: 'none of your sons is someone capable of being a judge'.

(5) Show the inconsistency of the clause set **symmetry** defined by:

```
(SETQ symmetry '( ((RELATION $X $X))
                  ((RELATION $U $W) (RELATION $U $V)
                                    (RELATION $W $V))
                  ((RELATION a b))
                  (() (RELATION b a))              ))
```

This set models the following problem: prove that every reflexive and semi-transitive relation is symmetrical. The first clause expresses reflexivity, the second semi-transitivity. The last two clauses come from negating the conjecture.

(6) Show the inconsistency of the clause set **concat1** defined by:

```
(SETQ concat1 '( ((APPEND nil $L $L))
                 ((APPEND (ht $U $X) $L (ht $U $Z)) (APPEND $X $L $Z))
                 (() (APPEND (ht 1 (ht 3 nil)) (ht 5 (ht 7 (ht 9 nil))) $X))
```

The first two clauses define the APPEND predicate: (APPEND $X $Y $Z) means that $Z is the result of concatenating the list $X (as the head of $Z) and the list $Y (as the tail of $Z); a similar definition was introduced in session 3.6 (§3.2.5) to encode APPEND in PROLOG. The notation (ht $X $Y) is used to represent a list whose head (the 'CAR') is $X and whose tail (the 'CDR') is $Y; the third clause is the result of negating the following conjecture: ∃ a list $x which is the concatenation of the list (ht 1 (ht 3 nil)) (as the head of the result) and the list (ht 5 (ht 7 (ht 9 nil))).

(7) Show the inconsistency of the clause set concat2 defined by:

```
(SETQ concat2 '( ((APPEND nil $L $L))
                 ((APPEND (ht $U $X) $L (ht $U $Z)) (APPEND $X $L $Z))
                 (() (APPEND $X
                             (ht 5 (ht 7 (ht 9 nil)))
                             (ht 1 (ht 3 (ht 5 (ht 7 (ht 9 nil)))))))
              ))
```

The first two clauses are exactly the same as those in the preceding example. The third is the consequence of negating the following conjecture: ∃ a list $x which, when concatenated (as the head) with the list (ht 5 (ht 7 (ht 9 nil))) gives the list (ht 1 (ht 3 (ht 5 (ht 7 (ht 9 nil))))).

These seven examples are now submitted to refuter-program 3.9. Comments follow session 3.16.

Session 3.16:

(hcsrbfpu)

YOUR COMMAND =>
OS

name of Old Set of clauses ? =>
set1

I can't find any justification

YOUR COMMAND =>
OS

name of Old Set of clauses ? =>
set2
Refutation succeeded

g6 = (p b) obtained by resolving
    g3 = (r a b) and literal 1 of
    g2 = (p b)or-(r a $y)
    substituting b for $y;

g5 = -(p $x) obtained by resolving
    g4 = (q $z3) and literal 2 of
    g1 = -(p $x)or-(q (f $x))
    substituting (f $x) for $z3;

g7 = () obtained by resolving
    g6 = (p b) and literal 1 of
    g5 = -(p $x)
    substituting b for $x;

YOUR COMMAND =>
OS
name of Old Set of clauses ? =>
set3
Refutation succeeded

g15 = (r c b) obtained by resolving
    g2 = (q c b) and literal 1 of
    g6 = (r $x $y)or-(q $x $y)
    substituting c for $x; b for $y;

g10 = (t e $z)or-(r c $z) obtained by resolving
    g4 = (q e c) and literal 1 of
    g8 = (t $x $z)or-(q $x $y)or-(r $y $z)
    substituting e for $x; c for $y;

g21 = (t e b) obtained by resolving
    g15 = (r c b) and literal 1 of
    g10 = (t e $z)or-(r c $z)
    substituting b for $z;

g28 = () obtained by resolving
    g21 = (t e b) and literal 1 of
    g9 = -(t e b)

YOUR COMMAND =>
OS
name of Old Set of clauses ? =>
logicians
Refutation succeeded

g6 = (sound a) obtained by resolving
    g5 = (judge a) and literal 1 of
    g3 = (sound $x)or-(judge $x)
    substituting a for $x;

g8 = (logician a) obtained by resolving
    g6 = (sound a) and literal 1 of
    g2 = (logician $x)or-(sound $x)
    substituting a for $x;

g7 = -(logician a) obtained by resolving
    g4 = (son a) and literal 1 of
    g1 = -(son $x)or-(logician $x)
    substituting a for $x;

g9 = () obtained by resolving
    g8 = (logician a) and literal 1 of
    g7 = -(logician a)

YOUR COMMAND =>
OS

name of Old Set of clauses ? =>
symmetry
Refutation succeeded

g6 = (relation $u a)or-(relation $u b) obtained by resolving
    g3 = (relation a b) and literal 2 of
    g2 = (relation $u $w)or-(relation $u $v)or-(relation $w $v)
    substituting a for $w; b for $v;

g15 = (relation b a) obtained by resolving
    g1 = (relation $x13 $x13) and literal 1 of
    g6 = (relation $u a)or-(relation $u b)
    substituting $u for $x13; b for $u;

g25 = () obtained by resolving
    g15 = (relation b a) and literal 1 of
    g4 = -(relation b a)

YOUR COMMAND =>
OS

name of Old Set of clauses ? =>
concat1

Refutation succeeded

g4 = (append (ht $u nil) $z (ht $u $z)) obtained by resolving
    g1 = (append nil $l2 $l2) and literal 1 of
    g2 = (append (ht $u $x) $l (ht $u $z))or-(append $x $l $z)
    substituting nil for $x; $l for $l2; $z for $l;

g5 = (append (ht $u (ht $u4 nil)) $l (ht $u (ht $u4 $l)))
obtained by resolving
    g4 = (append (ht $u4 nil) $z4 (ht $u4 $z4)) and literal 1 of
    g2 = (append (ht $u $x) $l (ht $u $z))or-(append $x $l $z)
    substituting (ht $u4 nil) for $x; $l for $z4; (ht $u4 $l) for $z;

g6 = () obtained by resolving
    g5 = (append (ht $u5 (ht $u9 nil)) $l5 (ht $u5 (ht $u9 $l5)))
and literal 1 of
    g3 = -(append (ht 1 (ht 3 nil)) (ht 5 (ht 7 (ht 9 nil))) $x)
    substituting 1 for $u5; 3 for $u9; (ht 5 (ht 7 (ht 9 nil))) for $l5; (ht 1 (ht 3 (ht 5
    (ht 7 (ht 9 nil))))) for $x;

(YOUR COMMAND) =>
OS

name of Old Set of clauses ? =>
concat2
Refutation succeeded

g4 = (append (ht $u nil) $z (ht $u $z)) obtained by resolving
    g1 = (append nil $l2 $l2) and literal 1 of
    g2 = (append (ht $u $x) $l (ht $u $z))or-(append $x $l $z)
    substituting nil for $x; $l for $l2; $z for $l;

g5 = (append (ht $u (ht $u4 nil)) $l (ht $u (ht $u4 $l)))
obtained by resolving
    g4 = (append (ht $u4 nil) $z4 (ht $u4 $z4)) and literal 1 of

g2 = (append (ht $u $x) $l (ht $u $z))or-(append $x $l $z)
   substituting (ht $u4 nil) for $x; $l for $z4; (ht $u4 $l) for $z;
g6 = () obtained by resolving
   g5 = (append (ht $u5 (ht $u9 nil)) $l5 (ht $u5 (ht $u9 $l5)))
and literal 1 of
   g3 = -(append $x (ht 5 (ht 7 (ht 9 nil))) (ht 1 (ht 3 (ht 5 (ht 7 (ht
9 nil))))))
   substituting (ht $u5 (ht $u9 nil)) for $x; (ht 5 (ht 7 (ht 9 nil))) for $l5; 1 for
$u5; 3 for $u9;

(YOUR COMMAND) =>
S
= bye

> If we were to write a PROLOG program whose first 3 instructions were the first 3 clauses of the **symmetry** set (example 5) in the same order and if the fourth clause were issued as a query, then the PROLOG program would loop *as a direct result of PROLOG's depth first strategy*. The reader can easily verify this with any PROLOG interpreter (including those developed at the beginning of the chapter). Conversely, the refuter-program presented here proves that **symmetry** is inconsistent regardless of the order of the clauses in the set. In fact, if S is a nondeterministic, refutation complete strategy, the deterministic strategy S_B obtained by driving S *breadth first* is complete.

> Let $S(E)$ be the *refutation graph* which can be reached from a clause set E obeying strategy S (the graph's nodes are the resolvents derivable from E, no matter how S is stipulated; there is an arc between each parent-clause and each resolvent-clause). If S_B finds an occurrence of the empty clause within the resolvents at level n in $S(E)$, it is guaranteed that the empty clause cannot occur at any level less than n in $S(E)$ (since, *in a breadth first strategy*, a resolvent at level $p+1$ cannot be derived until all the resolvents at levels less than or equal to p have been derived).

A necessary set of resolutions to prove that the empty clause appears permits a refutation graph to be formed, of the type presented in Fig. 3.3 (§ 3.3.3). If we impose a direction on the edges so that they go from each resolvent to their parent clauses and if we do not merge identical resolvents coming from distinct parent pairs, we get a *refutation arborescence* (for the input clause set) whose *root* is the empty clause and whose *leaves* are input clauses; the *depth of the arborescence* is defined as the number of arcs on the longest *path* from the root to any leaf. Given an inconsistent clause set it may be possible to extract different refutation arborescences, with varying depths, according to what strategies are followed. As a consequence of the property brought out in the preceding paragraph, guiding any nondeterministic, complete strategy S in a breadth first fashion gives a refutation graph of minimum depth compared with all those contained in $S(E)$.

For each example (2 to 7 above, plus the earlier examples of **sickpeople** and **identright1**) the table below shows the number of resolvents produced (i.e. the number of resolutions which *succeeded*, not to be confused with the number of resolutions *attempted*), the number of *useful* resolutions which led to the empty clause and the depth of the refutation arborescence associated with the solution. The difference between the first two numbers may be important; as indicated in § 3.3.6, in order to reduce this difference it would be possible to stop saving (and hence later on examining) resolvents that, apart from variable names, are identical to previously saved resolvents. This will be undertaken further on (see program 3.10).

clause set identifier	depth of the refutation arborescence	number of useful resolutions leading to the empty clause	number of resolvents produced
sickpeople	3	4	9
identright1	4	4	165
set2	2	3	3
set3	3	4	19
logicians	3	4	4
symmetry	3	3	21
concat1	3	3	3
concat2	3	3	3

> [CHANG & LEE 19 73] discusses how the "TPU" refuter deals with the identright1 example. This refuter is based on a strategy that is a hybrid of a unit strategy (TPU = Theorem Prover with Unit binary resolution; but it has a *non* positive unit strategy) and a 'set-of-support' strategy; the production of resolvents is driven *almost breadth first* (as soon as a level p unit resolvent appears an attempt is made to unify it with already existing unit clauses, before continuing to build level p, if necessary). Furthermore, the unit resolvents containing terms beyond a certain *level of nesting* (chosen by the user) are eliminated (the search graph is said to be *pruned*); for example, in (P a (f (g (h a b)))), the second argument's level of nesting is 3. The TPU refuter accepts any clauses (meaning not necessarily Horn clauses).

TPU's strategy is not complete, for several reasons: (i) the *binary* resolution unit strategy is complete for Horn clauses but not for any type of clauses in general, (ii) the set-of-support strategy is complete for any kind of clauses but completeness is lost when this strategy is combined with the unit strategy, and (iii) when resolvents beyond a certain level of nesting (arbitrarily fixed in advance) are eliminated, completeness can not be guaranteed. For the identright1 example and each one of the TPU and hcsrbfpu programs, the table below allows the depth of the refutation arborescence (depth), the number of useful resolutions (nur) and the number of resolutions produced (nrp) to be compared.

	TPU			hcsrbfpu		
	depth	nur	nrp	depth	nur	nrp
identright1	4	4	16	4	4	165

> It can be seen that the value of nrp is considerably higher under hcsrbfpu than under TPU. This advantage of TPU over hcsrbfpu stems partly from the elimination of unit clauses by 'subsumption'. A clause $\{Li\}_{i=1,p}$ is said to subsume a *clause* $\{Mj\}_{j=1,q}$ if there exists a substitution s such that $\{Li\}s$ is a subset of the set of literals $\{Mj\}$; conversely, $\{Mj\}_{j=1,q}$ is said to *be subsumed* by $\{Li\}_{i=1,p}$. For example, P(x) subsumes (P)y $\vee$ Q(z) as well as P(a), P(a) $\vee$ Q(z) and P(f(y)) $\vee$ Q(a,z); and P(x) $\vee$ Q(a) subsumes P(b) $\vee$ Q(a) $\vee$ R(y). Let E be a clause set and E' the set derived from E by eliminating the clauses subsumed by other clauses in E; it can be shown that 'E is inconsistent' is equivalent to 'E' is inconsistent'. In order to reduce the search space, but without eliminating all occurrences of the empty clause, it is a good idea to eliminate resolvents subsumed (by those already obtained) as they are produced. However, the subsumption test can itself turn out to be expensive. In TPU, this test is applied only to the unit resolvents.
> We should also stress that TPU's results depend on how judicious (or fortunate) the choice of the *nesting level limit*, which is made prior to each application, proves to be. If this level is too shallow, TPU will fail; if it is too deep, the development of the search graph will be less focused (a similar dilemma is faced when using a depth first search with a fixed depth limit).

We now propose a final version of the hcsrbfpu refuter (program 3.10) that eliminates any unit resolvent recognised to be subsumed by unit resolvents which have been already produced and saved. This version will be able to cope with examples (examples 8, 9 and 10 of session 3.18) which entail the formation of a prohibitively large number of resolvents when submitted to program 3.9.

3.3.12   Third version of the refuter (with elimination of subsumed unit clauses)

The definition of the subsumption of one clause by another was reviewed at the end of the preceding section. Program 3.10 only tests for, and makes use of, the subsumption of positive (or negative) unit clauses by other positive (or respectively negative) unit clauses. If one were to test for subsumption between any clauses one might anticipate discarding further resolvents (thus limiting the later resolutions), but at the price of a considerable additional effort; it would be interesting to evaluate the worth of such an effort experimentally.

Program 3.10 defined below comes from program 3.9 by modifying the procedures newlevel, collect, resolution and order, and by adding the procedures useful?, subsum+, subsum- and subsume. All these procedures are laid out below. New parts and new procedure names are underlined. Still with the objective of saving space, the procedures of program 3.9 which are unchanged are not reproduced; to get the whole of program 3.10 (whose use is illustrated in sessions 3.17 and 3.18) we retain the last versions of: hcsrbfpu, $, prompt, wipe, invoke, rename, renam1, singu, applysubstit, unification, len, disagreement, variable?, in?, tell, verse, writeresolvent, writeparents, ren, writemgu, tree, tree1, paste, pclear and pclear-.

Procedure subsume returns T or NIL according to whether the literal represented by the first argument (exp1) does or does not subsume the literal represented by the second argument (exp2); exp1 subsumes exp2 if and only if there is a substitution s which, when applied to the *variables of exp1*, gives exp2.

Note that in order to determine whether s exists, exp1 and exp2 are compared from a *pattern-matching* or *semi-unification* point of view (see chapter 2, §2.4.1). subsume can actually be derived by simplifying the procedure unification: it is sufficient to treat exp2 as having no variables, only symbols of constants and functions; to do this, instead of testing whether, say, the literal (P (* X) (* Y)) subsumes (P (g (* Y)) (* Z)), the test will be whether (P (* X) (* Y)) subsumes (P (g $Y) $Z). Recall that as far as variable? (which is called by subsume and by unification) is concerned a variable (after the $ procedure has acted on the input clauses) is a list whose first element is the * symbol; if, for example, the expression (g (* Y)) is replaced by (g $Y), variable? will no longer detect any variable. This is the reason why one of the literals passed as an argument to subsume (in this instance the one represented by exp2) is first submitted to procedure paste; this preprocessing is performed before every call to subsume, which means whenever useful? calls subsum+ or subsum− which, in turn, call subsume. In these conditions the renaming of variables and the 'occurs-check' test (see §3.1.1) are not relevant.

The useful? procedure's task is to determine whether the new resolvent NR (established by resolution) that is passed to it is a unit clause which is subsumed by another unit clause which has already been stored away. If NR is a *positive* unit clause, useful? calls subsum+ to test whether among the positive unit clauses previously saved (in theoldposu and in otheru+) there is one which subsumes NR. When one or other of the procedures subsum+ or subsum− succeeds in discovering that NR is subsumed, the escape resol (declared in resolution) is immediately performed: procedures subsum+ or subsum−, then useful? and resolution are abandoned; as a result, the final instructions in resolution which are intended to install the stored up resolvents (creation of an individual label, setting of the parents and mgu properties, ordering of the label — via order — into otheru+ or other≠u+ if the resolvent is not the empty clause) are skipped.

Note: from now on it is resolution, and no longer collect, that watches for the empty clause to appear; once this happens, the escape new (declared in newlevel) causes resolution, collect and newlevel to be exited, returning the empty clause's label to invoke.

Program 3.10

```
(DE newlevel () ;Called by invoke
;thenewposu, thenewnonposu, theoldposu, theoldnonposu are declared in
;hcsrbfpu and initialised to NIL on each call of init (in init)
;the escape new is used in resolution
    (TAG new
        (PROG (otheru+ other≠u+ allnonposu)
            (SETQ allnonposu (APPEND thenewnonposu theoldnonposu))
            (collect thenewposu allnonposu allnonposu)
            (collect theoldposu thenewnonposu thenewnonposu)
            (COND ((AND (NULL otheru+) (NULL other≠u+)) (RETURN
            'failure)))
            (SETQ theoldposu (APPEND thenewposu theoldposu)
                    thenewposu otheru+
                    theoldnonposu allnonposu thenewnonposu other≠u+)
            (RETURN NIL))))

(DE collect (theu+ the≠u+ refnonu) ;Called by newlevel
    (WHILE theu+
        (WHILE the=u+
            (FOR (i 1 1 (1- (LENGTH (EVAL (CAR the≠u+)))))
                (resolution (CAR theu+) (CAR the≠u+) i))
            (NEXTL the≠u+))
        (SETQ the≠u+ refnonu)
        (NEXTL theu+)))

(DE resolution (u+ nonu+ n) ;Called by collect
;the escape resol declared here is used by subsum+ and subsum-
;(which are called by useful?)
;the escape new is declared in newlevel
    (PROG (literal substitutions resolvent child ind)
        (SETQ literal (NTH n (EVAL nonu+))
                ind index
                substitutions (unification (rename (CAR (EVAL u+))) literal))
        (COND ((EQ 'failure substitutions) (RETURN NIL)))
        (SETQ resolvent (applysubstit substitutions (REMOVE literal (EVAL
        nonu+))))
        (TAG resol
            (COND ((NOT (EQUAL '(()) resolvent)) (useful? resolvent)))
            (SETQ child (GENSYM))
            (SET child resolvent)
            (PUTPROP child (LIST u+ nonu+ n ind) 'parents)
            (PUTPROP child substitutions 'mgu)
            (COND ((EQUAL '(()) resolvent) (EXIT new child))
                    (T (order child)))))))

(DE useful? (exp) ;Called by resolution
;otheru+ and other≠u+ are global, declared and initialised in newlevel
;theoldposu, theoldnonposu are global, declared in hcsrbfpu and
;updated in newlevel
```

```
      (COND (  (NULL (CDR exp))
               (SETQ exp (paste (CAR exp)))
               (subsum+ exp theoldposu)
               (subsum+ exp otheru+))
          (T (COND ( (AND (NULL (CAR exp)) (NULL (CDDR exp)))
                    (SETQ exp (paste (CADR exp)))
                    (subsum- exp theoldnonposu)
                    (subsum- exp other≠u+))))))

(DE order (resolvent) ;Called by resolution
;otheru+ and other≠u+ are global variables declared and initialised in
newlevel
      (COND ((NULL (CDR (EVAL resolvent))) (NEWL otheru+ resolvent))
            (T (NEWL other≠u+ resolvent))))

(DE subsum+ (exp list) ;Called by useful?
;the escape route resol is declared in resolution
      (MAPC '(LAMBDA (g) (COND ((subsume (CAR (EVAL g)) exp)
                                          (EXIT resol))))
              list))

(DE subsum- (exp list) ;Called by useful?
;the escape resol is declared in resolution
      (MAPC '(LAMBDA (x)
               (SETQ x (EVAL x))
               (COND ((AND (NULL (CAR x)) (NULL (CDDR x))
                                          (subsume (CADR x) exp))
                      (EXIT resol))))
              list))

(DE subsume (exp1 exp2) ;Called by subsum+ and subsum-
      (PROG (d)
        (COND ((NOT (EQ (len exp1) (len exp2))) (RETURN NIL)))
        (UNTIL (NULL (SETQ d (disagreement exp1 exp2)))
          (COND ((variable? (CAR d)) (SETQ exp1
                                (SUBST (CADR d) (CAR d) exp1)))
                (T (RETURN NIL))))
        (RETURN T)))
```

Session 3.17 below was obtained having loaded program 3.10; the
identifiers sickpeople, symmetry and identright1 had previously been
initialised as in the earlier sessions. The processing of these three examples
will give rise to some comments about the differences in behaviour between
versions 3.9 and 3.10 of the hcsrbfpu refuter. On the other hand, the way
version 3.10 treats the clause sets set2, set3, logicians, concat1 and
concat2 is not presented since the action of version 3.10 on each of these
examples is absolutely identical to that of version 3.9 (same resolvents
stored, hence same refutations and same values of depth, nur and nrp in
each case).

Session 3.17

(hcsrbfpu)

YOUR COMMAND =>

OS

name of Old Set of clauses ? =>
sickpeople
Refutation succeeded

g7 = (likes a b) obtained by resolving
  g4 = (doc b) and literal 1 of
  g2 = (likes a $x)or-(doc $x)
  substituting b for $x;

g8 = -(quack $y)or-(likes a $y) obtained by resolving
  g1 = (sick a) and literal 1 of
  g3 = -(sick $x)or-(quack $y)or-(likes $x $y)
  substituting a for $x;

g12 = -(likes a b) obtained by resolving
  g5 = (quack b) and literal 1 of
  g8 = -(quack $y)or-(likes a $y)
  substituting b for $y;

g13 = () obtained by resolving
  g7 = (likes a b) and literal 1 of
  g12 = -(likes a b)

YOUR COMMAND =>
OS

name of Old Set of clauses ? =>
symmetry
Refutation succeeded

g7 = (equal $v $w)or-(equal $w $v) obtained by resolving
  g1 = (equal $x5 $x5) and literal 1 of
  g2 = (equal $u $w)or-(equal $u $v)or-(equal $w $v)
  substituting $u for $x5; $v for $u;

g10 = (equal b a) obtained by resolving
  g3 = (equal a b) and literal 1 of
  g7 = (equal $v $w)or-(equal $w $v)
  substituting a for $w; b for $v;

g12 = () obtained by resolving
  g10 = (equal b a) and literal 1 of
  g4 = -(equal b a)

YOUR COMMAND =>
OS

name of Old Set of clauses ? =>
identright1
Refutation succeeded

g15 = (p $u $z $w)or-(p $y $z $v)or-(p (g $y $u) $v $w)
obtained by resolving
  g1 = (p (g $x11 $y11) $x11 $y11) and literal 1 of
  g4 = (p $u $z $w)or-(p $x $y $u)or-(p $y $z $v)or-(p $x $v $w)
  substituting (g $x11 $y11) for $x; $y for $x11; $u for $y11;

g45 = (p $w $z $w)or-(p $v $z $v) obtained by resolving
    g1 = (p (g $x44 $y44) $x44 $y44) and literal 2 of
    g15 = (p $u $z $w)or-(p $y $z $v)or-(p (g $y $u) $v $w)
    substituting $y for $x44; $u for $y44; $v for $y; $w for $u;

g67 = (p $w (h $v $v) $w) obtained by resolving
    g2 = (p $x79 (h $x79 $y79) $y79) and literal 1 of
    g45 = (p $w $z $w)or-(p $v $z $v)
    substituting $v for $x79; (h $v $y79) for $z; $v for $y79;

g189 = () obtained by resolving
    g67 = (p $w375 (h $v375 $v375) $w375) and literal 1 of
    g3 = -(p (k $x) $x (k $x))
    substituting (k $x) for $w375; (h $v375 $v375) for $x;

YOUR COMMAND =>
S
= bye

The table below compares the processing of the sickpeople, symmetry and identright1 examples by versions 3.9 and 3.10 of hcsrbfpu.

clause set identifier	hcsrbfpu version 3.9			hcsrbfpu version 3.10		
	depth	nrU	nrP	depth	nrU	nrP
sickpeople	3	4	9	3	4	8
symmetry	3	3	21	3	3	8
identright1	4	4	165	4	4	184

It can be seen that the depth of the refutation arborescence obtained by both versions is identical in all 3 cases (3, 3 and 4 respectively); generally speaking, the depth of the refutation arborescence contained by simple positive unit refutation can be predicted to be less than or equal to the depth of the refutation tree obtained by positive unit resolution *and* elimination of certain subsumed resolvents.

It can equally be seen that the refutation arborescence obtained by both versions contains the same number of non-leaf nodes (number of useful resolutions: nur) in all 3 cases; however, in each instance the refutation arborescences obtained by the individual versions are different (for the sickpeople example, compare with session 3.13; for the identright1 example, compare with session 3.15; for the symmetry example, compare with session 3.16).

Finally, it can be seen that in 2 out of the 3 cases version 3.10 saves fewer resolvents (look at nrp) than version 3.9. Nonetheless, the third situation is interesting; levels 1, 2 and 3 produced by program 3.10 are respectively included within levels 1, 2 and 3 produced by program 3.9; in fact, generally speaking, if programs 3.9 and 3.10 reach their respective $p+1$ levels, every level $q \leq p$ produced by program 3.10 is certain to be included in the same level produced by program 3.9. But if the two versions both discover an occurrence of the empty clause in their respective p+1 levels, the part of level p+1 which is then produced by program 3.9 can be of a lesser cardinality than the part of level p+1 produced by program 3.10. This is what happens (with p+1=4) in the case of identright1, hence the resulting values of nrp.

3.3.13 Further example applications

The examples dealt with in § 3.3.11 are completed by 3 new problems.

(1) Prove the inconsistency of the clause set inverseright defined by:

```
(SETQ inverseright '( ((p e $x $x))
                      ((p (i $x) $x e))
                      (() (p a $x e))
                      ((p $u $z $w) (p $x $y $u) (p $y $z $v) (p $x $v $w))
                      ((p $x $v $w) (p $x $y $u) (p $y $z $v) (p $u $z $w))
                    ))
```

This clause set models the following problem: given a set with an associative binary operation £, taking an identity element e on the left, such that every element has an inverse on the left in relation to e, prove that every element has an inverse on the right in relation to e.

The first clause describes the identity element, e, on the left; the second expresses the fact that, for every element, an inverse exists on the left in relation to e; the last two express the associativity of £; the third is the negation of the conjecture.

(2) Prove the inconsistency of the clause set commutative defined by:

```
(SETQ commutative '( ((p e $x $x))
                     ((p $x e $x))
                     ((p $x $x e))
                     ((p a b c))
                     (() (p b a c))
                     ((p $u $z $w) (p $x $y $u) (p $y $z $v) (p $x $v $w))
                     ((p $x $v $w) (p $x $y $u) (p $y $z $v) (p $u $z $w))
                   ))
```

This clause set models the following problem: given a set with an associative binary operation £, taking an identity element e, such that the square of every element is e, prove that the £ operation is commutative. The first two clauses describe an identity element e; the third clause states that the square of any element is e; the fourth and fifth come from the negation of the conjecture; the final two express the associativity of £.

(3) Prove the inconsistency of the clause set identright2 defined by:

```
(SETQ identright2 '( ((p e $x $x))
                     ((p (i $x) $x e))
                     (() (p a e a))
                     ((p $u $z $w) (p $x $y $u) (p $y $z $v) (p $x $v $w))
                     ((p $x $v $w) (p $x $y $u) (p $y $z $v) (p $u $z $w))
                   ))
```

This clause set models the following problem: given a set with an associative binary operation £, taking an identity element e on the left, such that every element has an inverse on the left in relation to e, prove that the £ operation takes e as the identity element on the right. 4 out of the 5 clauses of problem 1 are present; the third clause is the negation of the conjecture.

Session 3.18 was obtained by applying program 3.10 to the three problems represented by inverseright, commutative and identright2.

Session 3.18
(hcsrbfpu)

YOUR COMMAND => OS

name of Old Set of clauses ? =>
commutative
Refutation succeeded

g15 = (p $y $v $w)or−(p $y $z $v)or−(p e $z $w)
obtained by resolving
    g3 = (p $x8 $x8 e) and literal 1 of
    g7 = (p $x $v $w)or−(p $x $y $u)or−(p $y $z $v)or−(p $u $z $w)
    substituting $x for $x8; $y for $x; e for $u;

g185 = (p $y $v $w)or−(p $y $w $v) obtained by resolving
    g1 = (p e $x208 $y208) and literal 2 of
    g15 = (p $y $v $w)or−(p $y $z $v)or−(p e $z $w)
    substituting $z for $x208; $w for $z;

g193 = (p a c b) obtained by resolving
    g4 = (p a b c) and literal 1 of
    g185 = (p $y $v $w)or−(p $y $w $v)
    substituting a for $y; b for $w; c for $v;

g19 = (p $u $z $w)or−(p $x $z $u)or−(p $x e $w)
obtained by resolving
    g3 = (p $x12 $x12 e) and literal 2 of
    g6 = (p $u $z $w)or−(p $x $y $u)or−(p $y $z $v)or−(p $x $v $w)
    substituting $y for $x12; $z for $y; e for $v;

g134 = (p $u $z $w)or−(p $w $z $u) obtained by resolving
    g2 = (p $x152 e $x152) and literal 2 of
    g19 = (p $u $z $w)or−(p $x $z $u)or−(p $x e $w)
    substituting $x for $x152; $w for $x;

g232 = (p b c a) obtained by resolving
    g193 = (p a c b) and literal 1 of
    g134 = (p $u $z $w)or−(p $w $z $u)
    substituting a for $w; c for $z; b for $u;

g195 = (p c b a) obtained by resolving
    g4 = (p a b c) and literal 1 of

g134 = (p $u $z $w)or−(p $w $z $u)
substituting a for $w; b for $z; c for $u;

g11 = (p $x $v c)or−(p $x $y a)or−(p $y b $v)
obtained by resolving
    g4 = (p a b c) and literal 3 of
    g7 = (p $x $v $w)or−(p $x $y $u)or−(p $y $z $v)or−(p $u $z $w)
    substituting a for $u; b for $z; c for $w;

g223 = (p $x a c)or−(p $x c a) obtained by resolving
    g195 = (p c b a) and literal 2 of
    g11 = (p $x $v c)or−(p $x $y a)or−(p $y b $v)
    substituting c for $y; a for $v;

g275 = (p b a c) obtained by resolving
    g232 = (p b c a) and literal 1 of
    g223 = (p $x a c)or−(p $x c a)
    substituting b for $x;

g577 = () obtained by resolving
    g275 = (p b a c) and literal 1 of
    g5 = -(p b a c)

YOUR COMMAND => OS
OS

name of Old Set of clauses ? =>
inverseright
Refutation succeeded

g6 = (p (i $y) $v $w)or−(p $y $z $v)or−(p e $z $w)
obtained by resolving
    g2 = (p (i $x1) $x1 e) and literal 1 of
    g5 = (p $x $v $w)or−(p $x $y $u)or−(p $y $z $v)or−(p $u $z $w)
    substituting (i $x1) for $x; $y for $x1; e for $u;

g59 = (p (i $y) $v $w)or−(p $y $w $v) obtained by resolving
    g1 = (p e $x62 $x62) and literal 2 of
    g6 = (p (i $y) $v $w)or−(p $y $z $v)or−(p e $z $w)
    substituting $z for $x62; $w for $z;

g60 = (p (i (i $w)) e $w) obtained by resolving
    g2 = (p (i $x63) $x63 e) and literal 1 of
    g59 = (p (i $y) $v $w)or−(p $y $w $v)
    substituting (i $x63) for $y; $w for $x63; e for $v;

g11 = (p $u $z e)or−(p (i $v) $y $u)or−(p $y $z $v)
obtained by resolving
    g2 = (p (i $x6) $x6 e) and literal 3 of
    g4 = (p $u $z $w)or−(p $x $y $u)or−(p $y $z $v)or−(p $x $v $w)
    substituting (i $x6) for $x; $v for $x6; e for $w;

g50 = (p $u $v e)or−(p (i $v) e $u) obtained by resolving
    g1 = (p e $x52 $x52) and literal 2 of
    g11 = (p $u $z e)or−(p (i $v) $y $u)or−(p $y $z $v)
    substituting e for $y; $z for $x52; $v for $z;

g95 = (p $u (i $u) e) obtained by resolving
    g60 = (p (i (i $w302)) e $w302) and literal 1 of
    g50 = (p $u $v e)or−(p (i $v) e $u)
    substituting (i $w302) for $v; $u for $w302;

g179 = () obtained by resolving
    g95 = (p $u502 (i $u502) e) and literal 1 of
    g3 = −(p a $x e)
    substituting a for $u502; (i a) for $x;

YOUR COMMAND =>
OS

name of Old Set of clauses ? =>
identright2
Refutation succeeded

g6 = (p (i $$) £v £w)or−(p £y £z £v)or−(p e £z £w)
obtained by resolving
    g2 = (p (i £x1) $x1 e) and literal 1 of
    g5 = (p $x $v $w)or−(p $x $y $u)or−(p $y $z $v)or−(p $u $z $w)
    substituting (i $x1) for $x; $y for $x1; e for $u;

g59 = (p (i $y) $v $w)or−(p $y $w $v) obtained by resolving
    g1 = (p e $x62 $x62) and literal 2 of
    g6 = (p (i $y) $v $w)or−(p $y $z $v)or−(p e $z $w)
    substituting $z for $x62; $w for $z;

g60 = (p (i (i $w)) e $w) obtained by resolving
    g2 = (p (i $x63) $x63 e) and literal 1 of
    g59 = (p (i $y) $v $w)or−(p $y $w $v)
    substituting (i $x63) for $y; $w for $x63; e for $v;

g11 = (p $u $z e)or−(p (i $v) $y $u)or−(p $y $z $v)
obtained by resolving
    g2 = (p (i $x6) $x6 e) and literal 3 of
    g4 = (p $u $z $w)or−(p $x $y $u)or−(p $y $z $v)or−(p $x $v $w)
    substituting (i $x6) for $x; $v for $x6; e for $w;

g50 = (p $u $v e)or−(p (i $v) e $u) obtained by resolving
    g1 = (p e $x52 $x52) and literal 2 of
    g11 = (p $u $z e)or−(p (i $v) $y $u)or−(p $y $z $v)
    substituting e for $y; $z for $x52; $v for $z;

g95 = (p $u (i $u) e) obtained by resolving
    g60 = (p (i (i $w302)) e $w302) and literal 1 of
    g50 = (p $u $v e)or−(p (i $v) e $u)
    substituting (i $w302) for $v; $u for $w302;

g7 = (p $x e $w)or−(p $x (i $z) $u)or−(p $u $z $w)
obtained by resolving
    g2 = (p (i $x2) $x2 e) and literal 2 of
    g5 = (p $x $v $w)or−(p $x $y $u)or−(p $y $z $v)or−(p $u $z $w)
    substituting (i $x2) for $y; $z for $x2; e for $v;

g57 = (p $x e $w)or−(p $x (i $w) e) obtained by resolving
    g1 = (p e $x60 $x60) and literal 2 of
    g7 = (p $x e $w)or−(p $x (i $z) $u)or−(p $u $z $w)
    substituting e for $u; $z for $x60; $w for $z;

g151 = (p $w e $w) obtained by resolving
    g95 = (p $u432 (i $u432) e) and literal 1 of
    g57 = (p $x e $w)or−(p $x (i $w) e)
    substituting $x for $u432; $w for $x;

g1417 = () obtained by resolving
    g151 = (p $w6157 e $w6157) and literal 1 of
    g3 = -(p a e a)
    substituting a for $w6157

YOUR COMMAND =>
OS
= bye

The table below compares the depth of the refutation tree (depth), the number of useful resolutions (nur) and the number of resolvents produced (nrp) for the 4 examples identright1, inverseright, commutative and identright2 under the hcsrbfpu (final version 3.10) and TPU programs.

clause set indentifier	TPU			hcsrbfpu		
	depth	nur	nrp	depth	nur	nrp
identidentright1	4	4	16	4	4	184
inverseright	7	7	63	5	7	174
commutative	10	10	131	6	11	570
identright2	10	10	57	6	10	1412

It can be seen that, in each of the 4 examples, the depth of the refutation tree produced by hcsrbfpu is less than or equal to that of the tree produced by TPU while the number of useful resolutions (nur) are equal. It cannot be guaranteed that this will always be the case since the underlying strategies of hcsrbfpu and TPU create refutation graphs which are not generally comparable. In the examples given, the search executed by TPU appears to be deeper but narrower than that executed by hcsrbfpu.

Remember that TPU is a hybrid of a unit strategy (not constrained to be positive) with a set-of-support strategy (the sets of support can be chosen in varying degrees of appropriateness at the time of invocation) and with limited depth of nesting in the resolvents; however, in the 4 examples shown here, TPU took the nesting depth to be infinite. Recall finally that hcsrbfpu possesses a theoretical advantage: its strategy is complete, while this is not so in TPU (but TPU does work on non Horn clauses).

3.3.14 Directions for further investigation

3.3.14.1 Reducing useless operations

The *implementation* efficiency of the last version (program 3.10) of hcsrbfpu can be improved while still obeying the basic — complete — strategy used so far.

Renaming variables less often

It would, for example, be better to rename variables in an expression E1 that we are trying to unify with E2 only after verifying that the lengths of E1 and E2 as well as their first elements (predicate symbols) are equal; many renaming operations would thereby be avoided. This improvement would be equally appropriate to the PROLOG type refuters presented in the first part of the chapter.

Favouring resolutions between 2 unit clauses

As another example, two groups of non positive unit clauses could be maintained: those which are negative unit and those which are not; resolutions involving two unit clauses could then be favoured when starting to construct a new level. This process is likely to make the empty clause appear sooner (except in particular circumstances briefly analysed below) since the program will start to build the first level containing the empty clause in an order that will avoid producing all the non-empty resolvents.

We should however point out that in certain specific situations, "hcsrbfpu" can make several negative literals of a non positive unit clause disappear at the same time in the course of a single resolution; under these conditions, resolution with a non unit clause can lead to the empty clause before all resolutions between unit clauses have been performed.

This behaviour is a consequence of the call to **REMOVE** in the "resolution" procedure; the resolution which "hcsrbfpu" then carries out can be regarded as a non binary resolution (*full resolution*, or resolution with the addition of *factorisation*); but it may also be regarded as a sequence of several binary resolutions between the same positive unit clause and each in turn of the literals in the non positive unit clause; we then stay within the framework agreed so far: binary resolution only.

Avoiding certain useless resolutions

As a further example, a certain number of new resolutions and subsumption tests may be avoided by incorporating into the refuter a very simple addition (cheap in relation to the benefit) to supervise the order in which resolutions are performed. **TPU** uses this technique.

Take, for instance, two unit clauses $C1 = P$ and $C2 = Q$, and a non unit clause $C3 = R \lor \Box P \lor \Box Q$. $C4 = R \lor \Box Q$ can be formed from C1 and C3, then $C5 = R \lor \Box P$ from C2 and C3, then $C6 = R$ from C2 and C4, and finally $C7 = R$ from C1 and C5. Actually it is *useless to try to resolve C2 with C4*, for C4 comes from C1 and C3; now, the resolvent [[C3, 2, C1], 2, C2] is identical to $[[C3, 3, C2], 2, C1]$, where the notation $[Ci, p, Cj]$ stands for the resolvent of the pth literal of Ci and the *unit* clause Cj.

On a broader level, *let us suppose* that the resolvents $R1 = [[Ci, 1, Cj], 1, Ck]$ and $R2 = [[Ci, 2, Ck], 1, Cj]$ *are simulatneously defined* (to simplify the presentation, we have ensured that the 2 literals involved in the resolutions with Cj and Ck are put at the head of Ci). It can then be seen that R1 and R2 are identical, apart from variable names; so it is pointless to attempt to produce R2 if R1 has been produced.

To partially exploit this property, unit clauses can be numbered (starting from 1) in the order they appear (either input or through resolution) and each non unit clause can have the number of its parent unit clause associated with it (there will always be one such parent other than for input clauses: 0 is associated with these); then no attempt is made to resolve a non unit clause with associated number r and a unit clause with number t unless r t. To illustrate this mechanism, suppose that i, j and k are the numbers associated with clauses Ci (non unit) and Cj and Ck (unit) respectively, where $i \leqslant j \leqslant k$. The resolution $[Ci, 1, Cj]$ is permitted because $i \leqslant j$ and the number j is associated with the resolvent; given that $j \leqslant k$, $R1 = [[Ci, 1, Cj], 1, Ck]$ is produced; given that $i \leqslant k$, $[Ci, 2, Ck]$ is produced and k is assigned to it; but given that we do not have $k < j$, no attempt is made to perform: $[[Ci, 2, Ck], 1, Cj] = R2$.

Note that if R2 had been produced, the subsumption test would finally have discarded it, but only after many useless operations (renaming, unification, subsumption).

3.3.14.2   *Extracting more information about the possible refutations*
hcsrbfpu can be equipped with additional capacities, without interfering with its basic strategy. For instance, one might allow the search for a refutation to be reinvoked (automatically or interactively requested) after finding an occurrence of the empty clause so as to discover another refutation. Alternatively, one could try to adapt hcsrbfpu to make it display the substitutions undergone in the course of the refutation by the variables of certain input clauses; this is what happens for the variables present in the query in PROLOG type refuters (see the first part of the chapter); perhaps several sets of substitutions could be discovered, one after the other.

3.3.14.3   *Further restricting the resolution graph*
If we are willing to take the risk of using a non-complete strategy, it would be possible, for example, to prevent resolvents exceeding a nesting level limit, fixed on invocation, from being saved up and investigated (as TPU can do).

Conclusion

This book has been designed to extend and illustrate a traditional AI course.

Three fundamental themes which are taught in many AI courses have been examined in this work: i) problem solving by heuristic searching of state spaces, ii) expert system generators, and iii) problem solving using theorem provers. Basic algorithms, accompanied by course reviews and complementary material, have been presented for each topic; LISP programs have then been described fully, in successive stages, and used.

In the first section, the A and A* family of *ordered search* algorithms were studied in fair detail; the derived A^*_ε and $A^{\alpha*}_\varepsilon$ algorithms were then tackled. Variants of these algorithms which allow partial developments, pruning phases or a bidirectional search were not considered; nor were algorithms for searching *sub-problem spaces* (these sub-problem spaces may be regarded as a generalisation of state spaces: the search is represented by means of *hyper-graphs* rather than standard *graphs*).

The second chapter saw the (gradual) implementation of several kinds of expert system generators, intended either for diagnosis or for producing action plans; a relatively sophisticated pattern-matching system was also developed for generators accepting variables in rules. The collection of implemented programs and examples investigated is representative of the capacities (particularly those of inference and explanation) to be found in commercial expert system generators. However, in order to deal with real applications the most important requirement would be to enrich both the language for encoding knowledge and the software environment (with tools to aid knowledge acquisition and management) of the generators presented here.

In the final section two types of theorem provers, both manipulating Horn clauses, were progressively constructed. One adheres closely to a PROLOG-type strategy (incomplete); it is designed as a *logic programming* language and offers the principal functions of common PROLOG dialects. The other theorem prover follows a complete strategy; it is not equipped with programming primitives which are analogous to those possessed by the first one, but it does deal with refutation problems *expressed in a truly declarative fashion* (the ability to solve queries is independent of the order of the clauses). Many other kinds of first order predicate logic provers could now be considered: alternative refutation by resolution strategies, inference rules other than the resolution principle, strategies other than refutation, dealing with non-Horn clauses . . .

The study of the concepts and methods of AI without any relation to computer programming is a rather arid, sterile, perhaps even pointless enterprise. The intention of this book has been to aid the assimilation of certain domain principles in the course of concrete implementations and experiments. Programming style has been influenced more by educational intentions and intelligibility than by considerations of strict computational efficiency; variants and extensions have been neglected. Nonetheless, it is hoped that this book will encourage readers to make further use of AI techniques and to push back the current frontiers of AI.

Bibliography

ALTY J. L. and COOMBS M. J., *Expert sytems: concepts and examples.* NCC publications, 1984.

BANERJI R., *Artificial Intelligence: a theoretical approach.* North-Holland, 1980.

BARR A. and FEIGENBAUM E. A. (Eds), *The handbook of Artificial Intelligence.* Kaufmann, 1982.

BENCHIMOL G., LEVINE, P. and POMEROL J. C., *Systèmes-experts dans l'entreprise.* Hermès, 1986.

BONNET A, HATON J. P. and TRUONG N., *Systèmes experts: vers la maîtrise technique.* Interéditions, 1986.

BUNDY A., *Artificial Intelligence.* Edinburgh Unversity Press, 1981.

BROWNSTON L., FARRELL R., KANT E. and MARTIN N., Programming Expert Systems in OPS5. Addison-Wesley, 1985.

BUCHANAN B. G. and SHORTLIFFE E. H., Rule-based expert systems the MYCIN experiments of the Stanford Heuristic Programming project. Addison-Wesley, 1984.

CAYROL M., FARRENY H and PRADE H., Accès linguistique associatif à une base de données. Actes de la 2a Conf. Int. sobre bases de datos en humanidades y ciencias sociales. Madrid, June 1980.

CAYROL M., FARRENY H and PRADE H., Fuzzy pattern matching. *Kybernetes,* **11,** 103–116, 1982.

CAYROL M., *Le langage LISP.* CEPADUES-Editions, 1983.

CHAILLOUX J., DEVIN M., DUPONT F., HULLOT J. M., SERPETTE B. and VUILLEMIN J., *Le-Lisp Version 15.2: manuel de référence.* 2nd edn., INRIA, May 1986.

CHANG LEE C. L. and LEE R. C. T., Symbolic logic and mechanical theorem proving. Academic Press, 1973.

CHARNIAK E., RIESBECK C. K. AND Mc DERMOTT D. V., *Artificial Intelligence programming.* Lawrence Erlbaum Associates, 1980.

CHARNIAK E. and McDERMOTT, D. V., *An introduction to Artificial Intelligence.* Addison-Wesley, 1984.

CLARK K. L. and McCABE F. G., *MICRO-PROLOG: programmer en logique.* Eyrolles, 1985.

CLOCKSIN W. F. and MELLISH C. S., *Programming in PROLOG.* Springer-Verlag, 1981.

CONDILLAC M., *PROLOG: fondements et applications.* Dunod, 1986.

Index

T